The Earth's Crust

The Earth's Crust

by William H. Matthews III

illustrated with photographs

FRANKLIN WATTS | NEW YORK | LONDON

◄—A FIRST BOOK—►

In memory of
my grandparents
Anna Kilpatrick Fain and Clem F. Fain, Sr.

Frontispiece, courtesy of the Kentucky Department of Public Information.

SBN 531-00724-3

Library of Congress Catalog Card Number: 76-134367
Copyright © 1971 by Franklin Watts, Inc.
Printed in the United States of America

4 5

Contents

1.	Earth's Rocky Rind	3
2.	Air, Water, Land, and Life	4
3.	Minerals, the Ingredients of Rocks	9
4.	Rocks — Building Blocks of the Earth's Crust	12
5.	The Nature of the Crust	22
6.	The Crust As Seen from Space	28
7.	What Lies Beneath the Crust?	32
8.	Volcanoes: "Peepholes" in the Earth's Crust	37
9.	When the Crust Shakes	46
10.	The Crust Bends and Buckles	51
11.	Earth's Changing Crust	58
12.	A Crustal Jigsaw Puzzle	64
13.	Is the Oceanic Crust Spreading?	69
14.	The Crust Reveals Earth's History	75
15.	Tapping Earth's Treasure Chest	78
	Glossary	84
	Index	88

Interaction of the earth's crust and atmosphere causes rock weathering that produces fertile soil. (U.S. Department of Agriculture)

1. Earth's Rocky Rind

As we go about our daily lives, most of us pay little attention to what lies beneath our feet. For this reason the earth's crust — like most commonplace objects — is generally taken for granted. Yet our very existence depends upon the relatively thin rocky rind that covers our globe.

We might compare this thin veneer of crustal rocks to the skin of an apple. As we know, the apple's skin represents but a very small fraction of the apple's diameter. In the same way, the crust represents — at the very most — less than 40 miles of the 4,000 miles from the surface to the center of the earth. So, comparatively speaking, the earth's skin is considerably thinner than the skin of an apple.

Although the rocks of its crust make up but a very small percentage of our planet's overall content, they are of great importance to man. For example, these rocks provide the solid foundation upon which we live. Like the hard covering of ice on a frozen pond, the crust provides support and prevents us from sinking downward into the less solid rocks in the earth's interior. The crust is also the source of many valuable mineral resources. Such vital materials as coal, petroleum, salt, sulfur, iron, and copper are but a few of the treasures that man extracts from the earth's crust. More important, the weathering and breakdown of crustal rocks produce the life-giving substance that we know as soil.

The earth's crust is important for yet another reason. To *geologists*, scientists who specialize in the study of the earth, the crust is a giant history book. In it can be found evidence of ancient volcanic eruptions, traces of great mountains that were lifted up and worn away, and the remains of prehistoric animals that vanished from the earth many millions of years ago. In fact, almost all that we know about the earth and its history has been "read" from the layered rocks of earth's outer crust.

3

The crust, then, is much more than the dirt beneath our feet. Indeed, were it not for earth's rocky skin, you would not be reading this book now. All the physical materials in our daily lives can usually be traced back to the rocks and soil upon which we live.

2. Air, Water, Land, and Life

Although man depends upon the earth's crust for his existence, it actually forms but a small part of the earth's total composition. What is the makeup of planet Earth? Most of us would probably answer, "Rocks and soil," for the average person thinks in terms of the solid earth we walk on. But solids are not the only form of matter in and on our planet. Liquids and gases — like the water and air so necessary for life — are as much a part of earth as the solid surface upon which we live.

These three forms of matter — gases, liquids, and solids — are usually described in terms of three rather distinct zones or spheres. The life-giving envelope of gas that completely surrounds our planet is called the *atmosphere*. The waters of the earth make up the *hydrosphere*, and the solid rocky part of our planet is called the *lithosphere*. These three zones of matter are quite different in composition, but the boundaries between them are not always distinct. Instead, the zones continually interact with each other as air comes into contact with rock, rock with water, and water with air. Thus, although the crust is confined to the lithosphere, it is helpful in studying it to know something about the gaseous and liquid parts of the earth and their relation to the rocks.

4

Our atmosphere has been called an ocean of air. Extending for hundreds of miles above sea level, it consists mostly of nitrogen, oxygen, carbon dioxide, and small amounts of rare gases such as neon and argon. Also present is water vapor that has been picked up from the hydrosphere, and countless tiny dust-sized fragments of the lithosphere. This thick blanket of gas serves many purposes. It provides the air that we breathe, and it acts as an insulating blanket to keep us warm at night and to protect us from the sun's raging heat in the daytime. The air also serves as a protective shield to fend off meteorites that might otherwise bombard the earth's surface. When meteors enter the atmosphere, friction-produced heat causes them to burst into flame. Luckily for us, most of these so-called falling stars are burned up before they hit the surface of our planet.

Wherever air touches water they affect each other. Water may be evaporated and incorporated into the atmosphere as clouds. This water may later be released on the land as snow or rain which will weather and erode the rocks of the crust. Atmospheric *weathering* is an important geologic process that has affected the crust since the beginning of time.

Like air, water is necessary for the existence of life. Most of earth's water is found in the far-reaching universal sea that covers almost 71 percent of the face of the earth. This great blanket of salt water fills the earth's larger surface depressions to an average depth of some two and one-half miles. The hydrosphere also includes the fresh water found in streams and lakes as well as that in the ground.

The "water sphere," like the atmosphere, constantly interacts with the crust. The restless sea is forever wearing away the land and depositing sediment on the ocean floor. Streams steadily gnaw at their banks and deepen their channels by erosion. Water is also at work inside the crust. Groundwater slowly dissolves minerals from the buried rock for-

5

The rocks in Bryce Canyon National Park have been carved into many interesting shapes by atmospheric weathering. (U.S. National Park Service)

The restless sea erodes the coast in one area but may build up the shore elsewhere. (U.S. National Park Service)

A dinosaur skeleton is an example of how the biosphere can add to the lithosphere. Such fossils also provide information about the nature of prehistoric life. (Dr. C. M. Sternberg, National Museum of Canada)

mations through which it passes and may eventually form great underground caverns. In short, water — ably assisted by weathering — has been the major force that has been at work on the crust for hundreds of millions of years.

Thanks to the presence of air and water, our planet is populated by a wide variety of plants and animals. This great swarm of organisms makes up the *biosphere*. The biosphere is responsible for the air that we breathe, for coal and petroleum, and for the formation of many types of rocks.

The biosphere is just as important as earth's other spheres, and it also greatly affects the crust. The realm of life continually interacts with air, land, and water to become involved in a number of earth processes.

Plants and animals frequently modify the crustal rocks with which they come in contact. In addition, the record of prehistoric life preserved in fossils has provided valuable clues to the history of the earth.

Despite the fact that our lives depend on earth's air and water and that we are part of the biosphere, geologists are primarily concerned with the lithosphere. Here are the minerals that form the rocks that make up continents and ocean basins. Here, too, is the crust with its soils, metals, and other mineral resources so vital to mankind.

3. Minerals, the Ingredients of Rocks

When asked, "What is the earth made of?" most people reply, "Rocks." Others may say, "Rocks and soil," or perhaps, "Minerals and rocks." Generally speaking, all of these answers could be considered correct. Rocks *do* make up most of the solid earth, and soil is simply the end result of weathered rock. But the key word here is "*mineral.*" Minerals are the basic ingredients from which rock — and the earth's crust — are made.

The crust, then, is made of rocks. Regardless of their color, size, shape, or weight, all rocks have one thing in common: they are composed of minerals. And — like everything else in the world — minerals are made up of various combinations of natural substances called *elements*. Each element consists of tiny particles of matter called *atoms*, and the atoms of one element differ from the atoms of all other elements. Thus, each atom has its own distinctive weight, size, and chemical com-

9

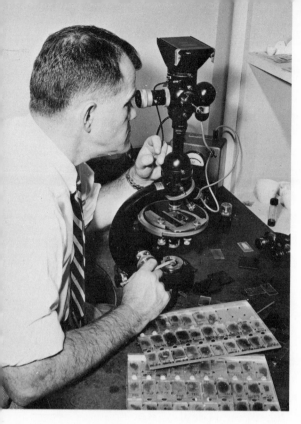

Geologists use special microscopes to deter-mine the type of minerals that are present in rocks. (U.S. Department of Agriculture)

position. Although some elements occur alone, they are commonly bound together to form a *compound*. Some minerals, such as copper, sulfur, gold, or diamond, may be composed of only one element. These are said to be in an *elemental,* or *native,* state. But most minerals are combinations of elements. Salt, for example, consists of the proper combination of the elements sodium and chlorine. Chemists refer to this chemical compound as sodium chloride. But the geologist knows it as the mineral called halite.

Despite the fact that chemists have identified more than one hundred elements, more than 98 percent of all minerals consist of only eight elements. These mineral-making elements are oxygen, silicon, aluminum, iron, calcium, sodium, potassium, and magnesium. Interestingly enough, more than three-fourths of the earth's crust is made of only *two* elements — oxygen and silicon.

Not all elements and compounds are minerals, of course. To be considered a mineral, a substance must occur in nature, possess certain distinctive physical characteristics, and have a definite chemical composition. The substance should also be *inorganic*. In other words, it should never have been part of, or derived from, any living thing. Finally, most minerals occur in definite shapes called *crystals*. These mineral crystals are helpful in distinguishing one mineral from another.

Mineralogists have recognized and named more than two thousand different minerals. These substances vary greatly in their chemical and physical properties, and these differences help to differentiate one mineral from another. But regardless of their makeup, minerals can generally be placed in one of two basic groups — the *metallic* and the *non-*

Each mineral has distinct physical and chemical properties. This piece of calcite displays double refraction — objects viewed through the mineral appear as a double image. (Ward's Natural Science Establishment)

metallic minerals. Most of us are familiar with the metallic, or *ore*, minerals. They yield the iron from which steel is made and are the source of such other widely used metals as lead, copper, tin, and gold. However, the less glamorous nonmetallic, or *industrial*, minerals are every bit as important. Salt is certainly not as exciting or as beautiful as gold. But, though we can live without gold, we must have salt to survive. Other nonmetals such as sulfur and gypsum are also widely used for industrial and medicinal purposes. Certain of the nonmetallic minerals form an especially important part of the earth's crust. These are the so-called rock-forming minerals. When chemically combined in the proper proportions, these minerals make rocks — the raw materials of the earth's crust.

4.   Rocks — Building Blocks of the Earth's Crust

The next time you take a hike, or even as you walk to school, look for rocks along the way. If you are observant, you will probably see a number of things that you have never noticed before. Chances are that you have kicked aside, or perhaps stumbled over, some interesting parts of the earth's crust.

Some people are so interested in rocks that they become avid rock collectors, or rock hounds. But like the air we breathe, the water we drink, and the ground beneath us, most of us simply take the presence of rocks for granted. We pick up a flat pebble to skim across a pond, step on a boulder to get across a stream, or use gravel or crushed stone

12

in our driveway. We seldom stop to think that glass is made from tiny grains of sand; brick and tile from clay; and many buildings are built of great blocks of sandstone and granite.

Rocks are not only all around us, but they can mean different things to different people. To the schoolboy, rocks are fun to throw or useful to build a fireplace on a camping trip. But the quarryman sees them in

Cores taken from oil wells provide petroleum geologists with information about crustal rocks of the subsurface. (Arabian American Oil Company)

a quite different light. Rock is his livelihood — a valuable product to be dug out of the crust and sold. The construction worker considers rock in a less friendly fashion. To him it is an obstacle that must be vigorously attacked with pick and shovel. Yet an engineer on the same construction site may welcome the presence of solid rock. He likes hard rock because it provides a firm foundation upon which to erect buildings and build roads.

The geologist looks at rocks from all of these viewpoints plus many, many more. Rocks are his stock-in-trade. He studies them in the laboratory and maps them in the field. He uses them to reconstruct the history of the earth and as a clue to the presence of valuable mineral resources. But more than anything else, the geologist considers rock in its most basic form — the stuff from which earth's crust is made.

If you collect very many rocks, they will probably be of varying shapes, sizes, and colors. This variety is not surprising, for rocks, like minerals, differ greatly in many respects. Different rocks are made of different minerals and it is the mineral content of the rocks that causes them to vary in weight, color, hardness, and composition.

Rock also differs in the way in which it is found in the earth's crust. As you walk along with your eye to the ground, you will note that many of the surface rocks are loose or *unconsolidated*. Soil, sand, gravel, and boulders are typical of this type of rock. Such loose material is known as the *mantlerock*. This blanket of rock debris is most important, for in the upper part of the weathered mantlerock, earth materials have become mixed with decaying plant and animal matter. This layer is the *soil*, that life-giving, earthen carpet that makes life possible.

Beneath this rocky blanket is the *bedrock* — a continuous mass of solid rock that has not yet been disturbed by surface agents such as weathering or erosion. Look for exposures of bedrock in the banks of a stream, the sides of a cliff, or where a road has been cut through a hill.

Mantlerock is the loose, unconsolidated blanket of rock debris that covers much of the earth's surface. (U.S. Department of Agriculture)

Bedrock was exposed here when glaciers eroded away the overlying mantlerock. (F. E. Matthes, U.S. Geological Survey)

Excavations for buildings and quarries are also good places to see bedrock.

Though mantlerock and bedrock differ in appearance, these two parts of the crust are actually quite closely related. In fact, one is the product of the other, for the mantlerock is composed of rock fragments produced by the weathering and erosion of once-solid bedrock. By the same token, the loose rock debris of the mantle may eventually become tightly packed, cemented together, and changed back into solid bedrock. The cementation, or *lithification*, of loose sand grains to form sandstone is an example of such a change.

If the mantlerock has been derived from crustal bedrock, where did

16

the bedrock come from? The major processes that form bedrock operate in two distinct geologic environments. One of these is found at or near the earth's surface. The other is located deep within, or even beneath, the crust. Thus, even though there are many different types of bedrock, all of them will have originated in one of these two rock-forming environments. This situation has made it possible for earth scientists to set up a system whereby rocks could be classified and described in terms of their origin. Although geologists may disagree on some points, they generally agree that most rocks can be fitted into one of three broad classes: *igneous*, *sedimentary*, and *metamorphic* rocks, the last being a result of a combined process.

The igneous rocks get their name from the Latin word *ignis*, which means "fire." This is a most appropriate name, for these rocks have been formed by the cooling and hardening of molten rock called *magma*.

These sedimentary rocks in the Badlands of South Dakota have been formed by the lithification of sediments. (U.S. Department of Agriculture)

Exposures of volcanic or extrusive igneous rocks are common in lava flows. (U.S. Geological Survey)

Some igneous rocks are produced by volcanic eruptions or lava flows. Such rocks are said to be *volcanic,* or *extrusive,* because they were extruded from within the crust and later hardened on the surface. Other igneous rocks, such as granite, hardened beneath the earth's surface. These are called *plutonic,* or *intrusive,* igneous rocks because they were intruded or injected into the buried rocks. In many parts of the crust, valuable deposits of metallic minerals (such as gold and silver) occur in association with intrusive rocks. Igneous rocks — especially intrusives

18

— make up a very large part of the crust. They become increasingly abundant at the lower depths, for 95 percent of the volume of the outermost ten miles of the crust consists of these "fire-formed" rocks. However, they are much less common on the earth's surface.

Because they cooled deep within the crust, intrusive rocks are typically seen in areas that have been greatly eroded. Over a period of millions of years, the intrusive rock masses have been gradually exposed as the overlying rocks have been stripped away by wind, water, and weathering. The granites on the bald summit of Colorado's Pikes Peak are a

Devils Tower in Wyoming is a good example of intrusive or plutonic rocks that have been exposed by erosion. (George A. Grant, U.S. National Park Service)

good example. Although now more than 14,000 feet above sea level, these intrusive rocks were undoubtedly formed many miles beneath the face of the earth.

In many places, the bedrock appears to consist of beds of rock stacked like the layers in a cake. This *bedding*, or *stratification*, is quite characteristic of the sedimentary rocks. These rocks were originally loose deposits of rock fragments such as sand, shells, mud, or gravel. But with the passage of time, these rock particles, called *sediments*, have hardened to form beds of sedimentary rocks. Most sediments, especially those derived from the breakdown of previously existing rocks, have been moved from their original location by some agent of erosion. The wind may pick up tiny dust-sized particles and carry them halfway around the world, and many rivers carry heavy loads of sand and silt. Glaciers can transport great quantities of rock debris ranging in size from tiny grains of sand to boulders the size of a house.

Sooner or later a geologic agent loses its ability to carry sediments and must deposit its load. When this happens, the sediments are typically laid down in layers. Known as *strata*, these layers or beds are the most distinctive feature of the sedimentary rocks. Such common rocks as sandstone, clay, and shale are typical examples of stratified sedimentary rocks.

Certain other types of sedimentary rocks may form as the result of chemical reactions. Gypsum and halite (rock salt) are two rather common nonmetallic mineral resources produced from chemicals that were once dissolved in water. Certain types of limestone have also been formed in this way. Still other sedimentary rocks consist of the remains or products of ancient plants and animals. Coal and certain kinds of fossil-bearing limestones are typical examples of rocks that originated in this manner.

If you have done much rock hunting, you have probably picked up a lot of sedimentary rocks. Although less than 5 percent of the outer ten miles of the crust consists of sedimentary rocks, they make up almost 75 percent of the rocks exposed on the earth's surface. This quantity is not surprising. Unlike the igneous rocks that originate underground, most sedimentary rocks form on or very near the earth's surface. These surface rocks not only weather to make soils, they are the source of many valuable products including water, coal, petroleum, salt, and other mineral resources. The sedimentary rocks are of particular interest to the geologists who study the earth's history. These layered rocks are stony pages of earth history which reveal much of our planet's past.

The metamorphic rocks are the third, and most complex, class of rocks. They get their name from two Greek words which literally mean "change in form." Metamorphic rocks can be formed from rocks that were originally igneous or sedimentary in origin. For example, the process of metamorphism may transform a fine-grained limestone into a much harder, crystalline, metamorphic rock called marble. This great change may take place when crustal rocks are subjected to intense heat and very great pressures. Such conditions typically come about during periods of great crustal deformation such as mountain building. They may also arise when hot intrusive rocks are injected into the cooler rocks that surround them. As rocks become metamorphosed, they normally undergo many changes in form and mineral composition. Thus, new minerals — some of great value — may form in rocks that were originally of no economic importance.

Although not as abundant as igneous or sedimentary rocks, metamorphic rocks generally have considerable geologic significance. Because they have had time to undergo such great change, metamorphic rocks are usually very old. And because they originate as a result of great heat, pressure, and crustal deformation, these "made-over" rocks provide information about some of the more violent chapters in the history of the earth.

Each type of rock, whether igneous, sedimentary, or metamorphic,

21

makes an important contribution to the overall composition of the crust. Equally significant, these different kinds of rocks provide the clues that help the geologist reconstruct the history of our planet and the many changes that it has undergone during the past 4½ billion years.

5.   The Nature of the Crust

Minerals are made of chemicals, rocks are made of minerals, and the crust is made of rocks. But how are these building blocks arranged? What is the true nature of this so-called crust?

Actually, the term "crust" is quite misleading, for it is all that remains of a now-abandoned theory of the earth's origin. For many years it was believed that most of the earth's interior consisted of hot, molten rock. The thin, outer, solid shell was thought to have formed when the upper part of the molten rock began to cool. As the melted rock cooled, it hardened, thereby producing a solid surface that formed as the molten rock crusted over. To early geologists this appeared to be a quite logical explanation. They knew that volcanoes spewed molten rock over the land, and this lava was thought to have come from the molten mass upon which earth's rocky rind was floating. Although geologists discarded this theory long ago, they still use the word "crust" to describe the outer rigid part of the lithosphere. But the use of this term in no way suggests the manner in which our planet was formed or of what its interior is made.

At this point you may be wondering how geologists know that our planet really *has* a crust. It would be easy to prove if the earth's crust was like that of a loaf of bread. In that case we could simply tear part of it away to see what was beneath it. But this is not the case; nor is the earth's crust like that of a pie. We can cut a slice of pie to see if it is filled with cherries or pumpkin, but our planet is just not sliceable!

Although man cannot cut or tear away the earth's crust, he has punched holes in it. Yet even the deepest of these — an unsuccessful oil test in west Texas — only penetrated the crust for a distance of about five miles. The deepest mine — and the deepest any man has ever ventured into the crust — is a two-mile-deep shaft in a South African gold mine. Needless to say, these punctures have barely penetrated earth's rocky skin.

However, surprisingly enough, scientists have learned a great deal about both the lower parts of the crust and the more deeply buried rocks which lie beneath it. Most of this information has been derived from the study of *seismic waves* generated by earthquakes. Whenever an earthquake occurs, these waves spread out in all directions from the earthquake's point of origin. This point, known as the *focus* or *hypocenter* of the quake, is typically located deep within the crust. When recorded by *seismographs* (sensitive instruments used to record the intensity of earth vibrations) the nature of the seismic waves can yield much information about the subsurface rocks.

Seismograms (the records made by a seismograph) indicate that there are three basic types of seismic waves. Each of these travels at a different *velocity* (speed) and reacts differently when passing through different types of matter. Thus, the behavior of these waves provides information about the type of material through which they are passing and their distance from the seismograph station.

The *compression*, or *push*, waves travel at velocities of from 3.4 to 8.6 miles per second. Because they are the fastest and the first waves to

23

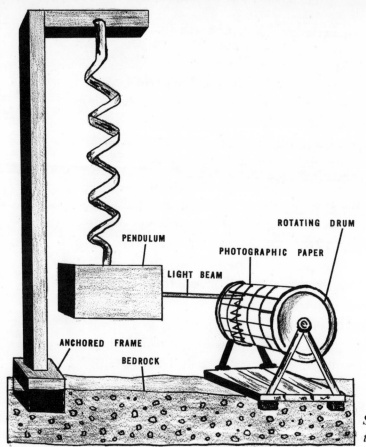

ROTATING DRUM

PENDULUM

PHOTOGRAPHIC PAPER

LIGHT BEAM

ANCHORED FRAME

BEDROCK

Seismographs record tremors in the earth's crust.

be recorded on the seismogram, they have also been called *primary*, or *P*, waves. The second set of waves to reach the seismograph are the *transverse*, or *shear*, waves. Known also as *secondary*, or *S*, waves, their speed ranges from 2.2 to 4.5 miles per second. Both P and S waves travel through the body of the earth and are thus called *body* waves.

The *L*, or *long*, wave, is the third and most destructive type of earthquake wave. These *surface* waves are generated from energy produced by the S and P waves, and they travel near the surface of the crust at speeds of from 2 to 2.5 miles per second. Because of their large *amplitude*, or size, they greatly disturb the surface rocks and cause most of

24

the actual earthquake damage. These waves originate at the earthquake's *epicenter* — the point on the earth's surface that lies directly over the hypocenter or focus.

Thanks to the record made by seismic waves we have a good idea about the nature of the crust and its underlying rocks as well. It has been learned, for example, that seismic waves travel much faster in very dense, or heavy, rocks than in lighter rocks. If one measures the time required for a wave to travel from its origin to a distant seismograph station, it is possible to estimate the *density*, or *specific gravity*, of the rock through which the wave has passed. *Seismologists* (earth scientists who specialize in the study of earthquakes) have also noted that earthquake waves are bent, or *deflected*, at certain levels within the earth. The depths at which the waves change direction also correspond to changes in the density of the rock at the depth where the waves are deflected.

These two characteristics — the velocity and deflection of seismic waves — made it possible to confirm the reality of the earth's crust. This important scientific discovery was made in 1909 by Professor Andrija Mohorovičić, a seismologist working in what is now Yugoslavia. While studying records from widely scattered seismograph stations, Mohorovičić noted striking similarities in the behavior of seismic waves at certain depths. Not only did these particular waves abruptly increase their speed, there was also a distinct change in their path of travel. Further study led Mohorovičić to conclude that the sudden change in behavior of these seismic waves was due to a corresponding change in the rocks through which they passed. In other words, the earthquake waves near the surface travel more slowly because the rocks of the crust are more rigid than those beneath them. The point at which the waves suddenly speeded up and changed course marks the base of the earth's crust. This transition zone has been named the *Mohorovičić discontinuity*, or *Moho*, in honor of the scientist who first recognized it. The discovery of the

25

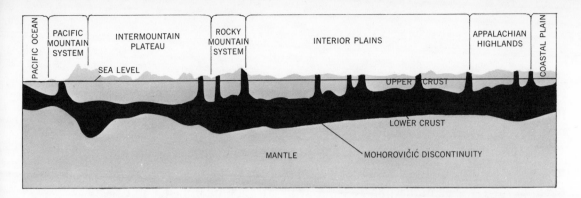

Cross section of mid-north America shows a schematic profile of the crust below the continent. The crust is much thinner below ocean basins; the continents are light crustal material floating on the mantle. (Reproduced by permission from Encyclopedia Americana, *Grolier Incorporated)*

Moho was a significant scientific breakthrough, for it provided geologists with a means of determining the thickness of the crustal rocks.

What do these speedy seismic messengers tell us about the composition and thickness of the crust? Despite the fact that crustal rocks make up only about one percent of the earth's volume, they have been found to differ greatly in specific gravity, thickness, and composition. Not only do the rocks vary at different depths in the crust, they also vary laterally or horizontally from one place to another. In other words, the crust is not a single, uniform, unchanging layer.

One of the more distinct crustal variations is the difference between the *oceanic crust* that makes up the ocean floors and the *continental crust* that forms earth's landmasses. Seismic studies suggest that the continental crust consists of two rather well-defined layers. There is an upper layer of dense, light-colored rocks similar in composition to granite. The granitic rocks of the upper crust are underlain by a lower layer of dark, slightly denser igneous rock called basalt. These basaltic rocks are similar to those typically produced by volcanic eruptions. The oce-

26

anic crust seems to be more uniform. It appears to be almost wholly composed of dense basaltic rocks similar to those of the lower part of the continental crust.

The continental and oceanic crusts also differ in their relative thickness and distribution over the face of the earth. Taken as a whole, the earth's crust is very thin when compared to the bulk of the earth. However, its thickness may vary greatly from one place to another. For example, it may be as much as 30 miles thick beneath a mountain range and as little as 3 miles thick in parts of the ocean basins. Beneath the continents, the crust averages about 25 miles in thickness. By contrast, the oceanic crust has an average thickness of about only 6 miles.

Why is there such a difference between the thickness of the oceanic crust and that of the continental crust? Information derived from a number of sources suggests that the crust literally floats on the heavier rocks beneath it, much as a ship floats on water. And since the crust is "floating," parts of it have sunk deeper into the earth than others. The continental crust is especially thick under great mountain ranges. In such areas the lower part of the crust has rootlike extensions which reach into the more plastic rocks below. Areas of lower elevation are also "floating." But because the crust is not so thick here, their "roots" do not extend as deep as the "roots" of mountains.

The density, or weight, of the rocks is also a factor. The continental crust is composed of relatively lightweight granitic rocks. Thus, their lighter weight permits them to "float" higher than the heavier basaltic rocks which make up the oceanic crust.

6. The Crust As Seen from Space

The date was July 20, 1969. On this historic evening, people the world around were glued to their television sets anxiously awaiting man's first step on the moon. Neil Armstrong not only made that "one small step for a man, one giant leap for mankind," he and his fellow Apollo astronauts returned safely to their home planet with a wealth of information and a priceless cargo of lunar samples. This record-breaking rock-collecting trip — the longest geological expedition that had ever been made by man — was a milestone in the history of science.

However, the benefits from the Apollo Mission and other space shots have not been confined to gathering information about the moon. Space vehicles are excellent observation platforms from which to photograph the earth and they have been used to good advantage. Taken from more than 240,000 miles in space, these photos have given earth scientists more than a birds-eye view of the earth's surface. Seen in this perspective it is possible to get some indication of the relative distribution of the oceanic and continental crusts.

Even a casual glance at these photographs reveals striking differences in the surface of the earth. Much of the surface appears to be relatively smooth. Elsewhere the earth's face is pimpled with rough, irregular areas. The smoother areas generally mark the presence of the *sea* — that great blanket of water that covers most of our globe. The rougher, uneven patches designate the high-standing parts of the crust. They are the *continents*, or *landmasses*, which cover about 40 percent of the earth's surface.

Rising to an average height of about 3 miles above the ocean floors, the continents are composed largely of granite. However, we can only see about 57.5 million square miles, or roughly 29 percent, of the con-

28

Astronaut Edwin Aldrin is seen deploying scientific experiment equipment on the moon. Information derived from these instruments may provide clues as to the origin of the earth's crust (National Aeronautics and Space Administration)

tinental masses from space. The remaining 11 percent of the continents consists of the *continental shelves*. These gently sloping extensions of the landmasses are actually the margins of the continents that are now covered by the sea.

Photographs taken nearer the earth reveal considerable detail about the uneven continental surface of the crust. In places there are flat, coastal plains which dip gently into the adjacent sea. At the other extreme are the great mountain systems such as the Rocky Mountains, the Alps, and the Himalayas. The highest point on the continents — Mount Everest — rises 29,028 feet above sea level. But most of the continental surface is located somewhere between these two extremes. On the average, the surface of the land rises only about one-half mile above sea level.

Space photos show that the bumpy landmasses make up but a relatively small part of the earth's surface. The smooth areas — which represent the sea — are much more widespread. Most of this water is located in earth's Southern Hemisphere. This has been called the water hemisphere, for 81 percent of this part of the crust lies beneath the sea. In space photos taken from above the North Pole, our planet looks quite different. More than 40 percent of the Northern Hemisphere lies above sea level. The overall view of both the Northern and Southern hemispheres reveals that almost 71 percent of the earth's surface is covered by the sea. This is a total of some 139 million square miles as opposed to the 57.5 million square miles of land that rises above the hydrosphere.

Although the part of the earth covered by the sea appears relatively smooth from space, this is quite deceptive, for the crust beneath the sea is far from smooth. Within recent years there has been much research into the nature of the sea floor. Oceanographic vessels have made rather detailed maps of certain ocean bottoms and collected countless samples of bottom sediments. Core holes have also been drilled through the sedi-

30

ment in an attempt to learn more about the structure and composition of the oceanic crust. Even so, much remains to be learned, for only about 6 percent of this part of the crust has been mapped with any accuracy. In view of this, it is not surprising that it has been said that we know more about the moon than we know about the surface of the earth.

Maps of the ocean floor reveal that the land beneath the sea is far from being flat and featureless. It is marked instead by rugged underwater mountain chains and deep submarine canyons and trenches. There is the Mid-Atlantic Ridge, a great submarine mountain system that zigzags its way for about 10,000 miles from the southern tip of Africa all the way to Iceland. In places, the rocky spine of this ridge rises almost a mile above the ocean bottom and it averages about 20 miles in width.

The floor of the sea is also slashed by yawning chasms that dwarf even

A great variety of submarine landforms are present on the surface of the ocean floor. (U.S. Naval Oceanographic Office)

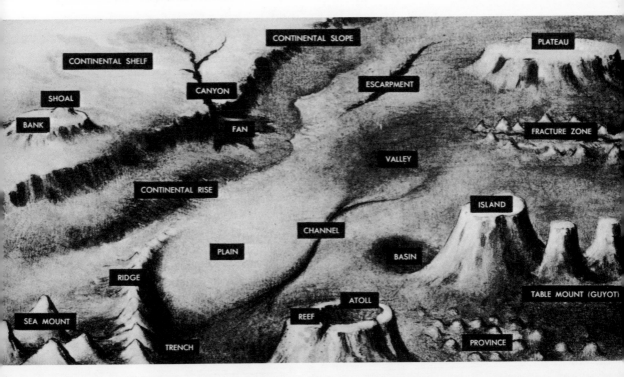

the famous Grand Canyon of Arizona. In fact, the deepest of these — the Mindanao Trench in the Pacific Ocean — is about seven times deeper than the Grand Canyon. At its deepest part it lies some 37,700 feet below sea level. The Puerto Rico Trench, in the depths of the Atlantic Ocean, plunges some 30,246 feet beneath the surface of the sea. The oceanic ridges and trenches are, of course, the extremes. In between, there are plains, plateaus and smaller depressions, and elevations of all shapes and sizes — just as there are on the land.

When referring to the physical features or the configuration of the earth's surface, the geologist speaks of the *topography* of an area. Thus, the topography and *relief* (the difference in elevation between the high and low points) of the ocean floor resemble those of the continents in many ways. Considered together, the continental landmasses and ocean basins make up the major *relief features* on our globe. The total relief of the crustal surface — from the summit of Mount Everest to the bottom of the Mindanao Trench — is more than 12 miles. Yet when considered in comparison with the great size of the earth, this range of elevation is hardly noticeable. In fact, on a model earth of 25 feet in diameter, mighty Mount Everest would rise less than one-quarter inch above sea level!

7.   What Lies Beneath the Crust?

Have you ever wondered what is inside the earth? In some places we drill wells and strike oil. Elsewhere a well may find pure water. One mine may yield gold; another may produce salt. Does this mean that our globe is filled with scattered pools of oil and water mingled with pockets of gold and salt? Or perhaps it is partly hollow and filled with air like some gigantic basketball. There are great hollow places in the earth and some of these caves are very large and deep. In fact, if you have been in a cave you have actually been inside the earth's crust — but not very far. Even at the deepest part of New Mexico's Carlsbad Caverns you were less than 800 feet underground. This is hardly a scratch on the total thickness of the crustal rocks.

If you have thought about inner space — as the earth's interior has been called — this is a perfectly normal reaction. Written history indicates that man has long been curious about the origin and general make-up of the earth. Early man's curiosity was not limited to the crust on which he walked. He also speculated on what might lie deep within this mysterious planet. Part of this interest was prompted by man's natural curiosity or need to know. However, much of it was caused by fear. Why, he wondered, did certain mountains occasionally explode and belch forth fire and suffocating clouds of smoke? And what unknown forces caused the ground to shake and giant sea waves to smash into the coast? Ancient man had no way of knowing what caused volcanic eruptions, earthquakes, and seismic sea waves. Indeed, we do not fully understand their cause today. Certain early thinkers did, nevertheless, associate these natural phenomena with disturbances inside the earth.

Lacking scientific knowledge about our restless planet, the ancients blamed these frightening events on evil or supernatural forces. Consequently, they attempted to explain them in terms of legends and myths. It was suggested, for example, that the earth was full of fire. When this fire escaped through the sides or top of a mountain, a volcano was formed. There was also a theory that the earth was filled with great pockets of air and water. Cracks in the crust permitted the trapped air to reach the surface in some places. Pressures generated when this air escaped produced the vibrations known as earthquakes. By the same token, great floods were thought to be caused when the underground waters burst out of the crust. It was even suggested that the earth was completely hollow except for its outer, rigid shell.

Although much has been learned about the earth since those ancient times, scientists must still theorize about its interior. We have explored the moon, photographed other planets from spacecraft, and are gradually solving the mysteries of the deep ocean floors. But it is most unlikely that man will ever venture more than a few miles into the crust of the earth. In the first place, finding a way to drill almost 4,000 miles into the earth presents engineers with a seemingly impossible task. Even if such a hole *could* be drilled, it is even more unlikely that man could use it for a journey to the center of the earth. There is a steady rise in temperature and pressure as one goes deeper into the earth. This increase continues until temperatures rise to as much as 4,000 to 5,000 degrees Fahrenheit, or almost half as hot as the surface of the sun. The pressure? At the earth's center, the pressure is approximately 3½ million times our atmospheric pressure here on the surface! Needless to say, it would require a most exceptional vehicle to protect an inner-space explorer from such incredible conditions.

Even though geologists cannot sample or study the inner earth directly, they still have a good idea of what it is like. Some information has been gathered from indirect evidence such as field and laboratory studies of certain rocks and minerals that occur in the upper crust. Investigations have also been made into the nature of earth's magnetic field, gravitational attraction, and the flow of heat from within the earth. These rock properties all have some relation to the overall structure of the earth. Clues have even come from outer space. The study of meteorites — which might be fragments of another planet — and the lunar

34

The mineral fragments found in these rocks collected in Antarctica suggest that they may be fragments of the earth's mantle. They were carried to the surface by magma. (National Science Foundation)

samples recovered by Apollo astronauts has also yielded bits of helpful information.

Nevertheless, most of what we know about inner space has come from the study of seismic waves. Upon leaving the crust and entering the Moho, the speeds of P and S waves are suddenly increased. This marks the top of the *mantle*, the second major zone in the earth's in-

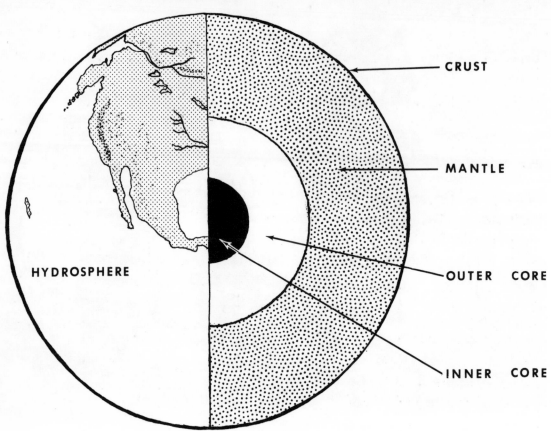

ATMOSPHERE

CRUST

MANTLE

OUTER CORE

INNER CORE

HYDROSPHERE

Cross section of the earth showing arrangement of the interior.

terior. The behavior of the seismic waves as they enter the mantle clearly indicates an abrupt change in the material through which they are passing. There is no direct evidence as to the mantle's composition, but it is generally believed to consist of very dense, solid rock. This 1,800-mile-thick layer represents more than 80 percent of the earth's total volume and appears to increase in density with depth.

Upon reaching the bottom of the mantle, earthquake waves give evidence of yet another — and more drastic change — as they enter the earth's *core*. The core is about 4,350 miles in diameter and is completely enclosed by the mantle. The behavior of seismic waves suggests that the core consists of two parts. There is an *outer core* that is about 1,360 miles thick and apparently in a liquid condition. This layer of fluid rock material completely surrounds the very dense, solid inner core.

Why is the outer core thought to be liquid? We again rely on seismic waves to provide the answer. As P waves travel through the outer core their speed is greatly reduced and they act as if they were passing through a liquid. However, upon reaching the *inner core* they suddenly speed up and behave as if they were traveling through a solid. More important, the S waves — which will not travel through a liquid — do not pass through the core at all.

The reaction of P waves has also provided some clues as to the chemical composition of the core. There is evidence to suggest that it is made up of from 80 to 85 percent iron with varying amounts of elements such as nickel, silicon, and cobalt. Interestingly enough, this is similar to the composition of certain meteorites which fall on the earth from space.

8.   Volcanoes: "Peepholes" in the Earth's Crust

May 8, 1902, dawned bright and sunny on the West Indies island of Martinique. In St. Pierre, the capital city, there was a great deal of excitement. Part of the hustle and bustle stemmed from the big election that was to be held on May 10. And the fact that the governor and his wife were in town added to it. Yet, there was also an undercurrent of fear among the good people of St. Pierre.

Indeed, there had been an uneasy feeling in the city since the unusual events of April 23. It was on that day that the streets of the city had unexpectedly been sprinkled with ashes and the air invaded by the sharp odor of sulfur. The startled residents did not have to look far to find the cause of their concern. Mount Pelée, a 4,500-foot volcanic mountain overlooking the city, appeared to be waking up after a long period of inactivity.

At first there appeared to be no cause for alarm, as Pelée's last known eruption had occurred in 1851. That outburst had lasted but a short time and did so little damage that it was soon forgotten. But this present disturbance seemed to be much more serious. On April 25, a series of explosions had blasted rocks, ashes, and steam out of one of the volcano's craters. And from that date until May 8 the outbursts became increasingly violent and more frequent. As time passed, the streets of St. Pierre became clogged by the continuous fall of ash and cinders, and many business houses had to close their doors. Meanwhile, poisonous gases filled the air and caused sore throats and eye irritations among the residents, and asphyxiated a number of animals.

Yet despite the preliminary eruptions, no one was quite prepared for the events of May 8. Just before eight o'clock on the fateful morning, Mount Pelée literally exploded. Four violent outbursts filled the air with hot clouds of steam, dust, and poisonous gases. Much of this volcanic debris soared skyward from the volcano's crater. But part of this deadly material boiled out of the side of old Pelée. Unfortunately, St. Pierre lay on this side of the mountain and directly in the path of this hurricanelike blast. Rushing down the mountainside at speeds of more than 100 miles per hour, it took only two minutes for this fiery cloud to completely engulf the city and destroy all that lay in its path. When the dust had

38

settled and the poisonous smoke had cleared away, an eight-square-mile area had been destroyed and more than 30,000 people had lost their lives.

What causes a mountain to blow up suddenly? Why, after almost a half of a century, should Mount Pelée bring about such death and destruction? This question has been asked for many centuries, for man has always viewed these "fire-breathing mountains" with awe and fear. Those early men who thought the earth was filled with fire said that volcanoes formed when flames burned through the crust. The early Romans had another idea. They believed that one of their "burning mountains" was the home of Vulcan, the Roman god of fire. Vulcan was also thought to be the blacksmith for the other gods. The noise and vibrations of the volcano's explosions were said to be caused by Vulcan pounding on his anvil. The smoke and ashes pouring from the crater were assumed to have come from the blacksmith's forge. It is interesting to note that the word "volcano" is derived from the Latin word *Volcanus* or *Vulcanus*. This was the name of the Italian volcano where the fire god was supposed to have lived.

Because of their effect on man, geologists have studied volcanoes in great detail. Fortunately, much has been learned about the way in which they erupt and the types of rocks that they produce. But much more remains to be learned, for the answers to the more puzzling volcanic mysteries remain locked up in the deeper rocks of the crust.

As noted earlier, volcanic rocks are the product of the cooling and hardening of molten rock material called magma. These "fire-formed" rocks poured from inside the crust as lava, and later cooled and hardened on the earth's surface. We can easily account for the occurrence of volcanic rocks, for they are forming in many parts of the world today. But the presence of pockets of melted rock in the solid crust is more difficult to explain.

Research has shown that magma occurs in isolated pockets, known

39

Lava from Parícutin volcano advances on a Mexican village. (Field Museum of Natural History)

as *magma reservoirs*. These chambers of melted rock originate no more than about 20 miles beneath the surface in the lower part of the crust or possibly in the upper mantle. Thus, there can be no direct connection between the magma reservoirs and earth's outer liquid core. There is not, as was once suggested, an underground "pipeline" leading from the core to the surface. However, the source of the heat that melts the crustal rocks and triggers volcanic eruptions is less clearly understood.

Volcanologists (geologists who specialize in the study of volcanoes) have given much thought to the problem of magma and have developed several theories as to how it may form. They all agree that extremely

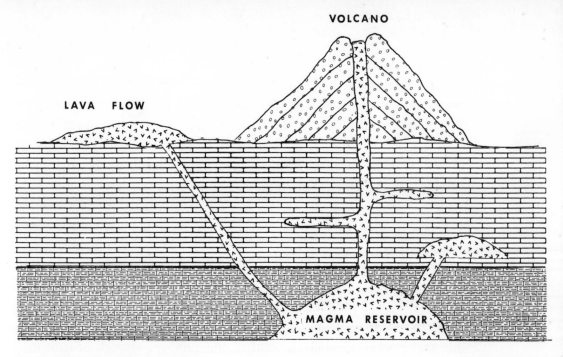

VOLCANO

LAVA FLOW

MAGMA RESERVOIR

Cross section of a typical volcano.

high temperatures are necessary to convert solid rock to the liquid state. It is further agreed that magma must form at depths where rocks are subject to tremendous pressure and intense heat. It is known that temperatures and pressures increase steadily toward the center of the earth. Could this account for the rocks melting at depth? Not according to our helpful seismic messengers, the P and S waves. Information from these earthquake waves shows that with the exception of isolated zones of igneous activity, most rocks in the crust and upper mantle are in the solid state. However, in those places where breaks occur in the earth's crust, the subsurface rocks may move toward the surface. As they rise upward, some of them may change to the liquid state as the result of diminishing pressure.

41

Some geologists think that the interior heat may be produced by friction. Whenever the earth's crust is broken or deformed by internal movements, great blocks of the crust are ground together and slide over one another. Much heat is generated as these rock masses come in contact, and this friction-produced heat might result in the melting of segments of the crust. This theory is supported by the fact that most areas of volcanic activity are associated with regions where there has been recent crustal deformation such as mountain building and earthquake activity.

It has also been suggested that the necessary heat may be given off by localized concentrations of radioactive elements. Minerals containing these energy-producing elements are abundant in certain crustal rocks, and some of them are capable of generating great heat. It is possible that pockets of such minerals could produce enough heat to melt the rocks enclosing them. As the rocks melted they would expand, perhaps exerting sufficient pressure to fracture the rocks above them. Magma could invade these fractures which might then provide zones of weakness to allow the molten rock to reach the surface.

The above explanations are based on theory rather than on facts. So the search for the true cause of volcanism continues. Meanwhile, most volcanologists think that a combination of factors produce rock-melting temperatures in the crust. Some of the heat is the result of the high temperatures that are naturally present in the deeper parts of the crust. Additional heat is derived from radioactive elements, with occasional help from heat generated during great crustal disturbances.

The rock material thrown out during a volcanic eruption is not lost to the crust. Instead, it is transferred from within the crust to become part of the surface rocks. This volcanic debris — lava, ash, cinders, and dust — may also alter the landscape. Great cone-shaped, volcanic mountains dot the land in many parts of the globe and are among the most

42

beautiful mountains in the world. Mount Rainier in Washington, Mount Shasta in California, and Japan's Mount Fujiyama are typical examples. Similar cones may build up from the ocean floor. Indeed, the Hawaiian Islands are part of a great chain of volcanic mountains that rise many thousands of feet above the bottom of the Pacific Ocean.

Volcanic activity does not always produce the familiar mound- or cone-shaped mountain. In some places, the lava has flowed out of the crust and spread over the land to form thick layers. Over a long period of time these lava flows may eventually become stacked on top of each

The Columbia River Plateau of the Pacific northwest is composed of great sheetlike lava flows resting one on top of the other. (T. O. Jones, U.S. Geological Survey)

other and cover very great areas. This is what happened in the Columbia River Plateau of Washington, Oregon, and Idaho. Here there are great sheetlike masses of lava that cover an area of about 200,000 square miles. This great lava plateau consists of hundreds of individual lava flows superimposed one on top of the other. Individual eruptions laid down separate layers of lava ranging from 10 to 15 feet thick. In places their total thickness is as much as 4,000 feet — a most substantial addition to the outer crust.

The study of volcanoes has yielded a wealth of information about the earth's crust. Not only do volcanoes account for the presence of many igneous rocks, they also give clues to physical conditions in the deeper parts of the earth. This has prompted one geologist to call volcanoes peepholes or windows in the earth's crust. One of these peepholes, Kilauea Volcano in Hawaii, is kept under constant observation by the United States Geological Survey. This is a very active volcano and the studies made here have been very helpful in the attempt to devise a method for predicting volcanic eruptions. Scientists assigned to the Hawaii Volcano Observatory also sample the lava and gases coming from the volcano. These materials can provide some indication of the nature of the deeper crustal rocks where they might have originated.

Volcanoes also tell us something about the structure of the crust. They indicate that there are certain places where the crustal rocks are weak and subject to fracture. In such areas the magma may force its way upward through great cracks or literally burn its way to the surface. These zones of weakness make up the world's *volcanic belts*, or *zones*. These areas coincide with the earth's major *seismic belts*, or zones of earthquake activity.

Where are these zones of crustal weakness located? Of the more than five hundred known active land volcanoes, most are located along the margin of the Pacific Ocean. Many more are probably present on the

44

Major volcanoes around the Pacific Ring of Fire.

ocean floor. Known as the Pacific Ring of Fire, this circle of volcanic activity includes the volcanoes of the East Indies, the Philippines, Japan, Alaska, and South and Central America.

The volcanoes of the Azores, West Indies, Turkey, India, Hawaii, and the Mediterranean make up the second major volcanic belt. This zone circles the earth approximately parallel to the equator. These two belts apparently indicate major zones of crustal weakness, but volcanoes also occur in the Antarctic and the Atlantic, Pacific, and Indian oceans.

9.   When the Crust Shakes

It was March 27, 1964, Good Friday. The place: Anchorage, Alaska. The residents of the city and of other communities in south-central Alaska were preparing for a quiet Easter weekend. Then it happened. At precisely 5:36 P.M., the ground shuddered violently and buildings began to sway and topple. Great gaping cracks opened up in streets and yards, triggering landslides that destroyed schools, homes, and churches. The tremors lasted only four minutes. But during those 240 seconds, more than 114 people lost their lives, and $750,000,000 worth of property was destroyed.

The results of this frightening disturbance were not confined to Alaska. Sudden displacement of the sea floor generated ocean waves that quickly spread across the Pacific. One of these killer waves struck Crescent City, California, leaving behind a trail of death and destruction. Another wall of water smashed into the Oregon coast drowning a

View of downtown Anchorage, Alaska, after the Good Friday earthquake of 1964. (Environmental Science Services Administration)

family of four. Similar waves were reported from such distant places as Chile, Japan, Hawaii, and Antarctica.

What powerful forces were responsible for the great crustal disruption that dealt such widespread damage? Anchorage had been rocked by one of the most severe earthquakes within recent time.

Few natural phenomena are as frightening and destructive as a major earthquake, and man has feared them since the dawn of time. He has

47

also wondered what mysterious underground forces could cause such jolting shocks. Certain of the early Greeks and Romans said the earth was filled with gas. They thought earthquakes were caused by pressures produced when the trapped air escaped to the earth's surface. Others said the sudden shocks occurred when the ceilings of large caverns collapsed and fell to the bottom of the cave. The Hindus had their own ideas about earthquakes. They believed that our globe rested on the back of an elephant that stood on a huge turtle. The turtle was standing on a coiled cobra; and when any of these animals moved, they caused the earth to shake. The lamas of Mongolia had a different explanation — they thought the earth rested on a frog. When the frog croaked or shifted, it caused the earth to vibrate.

Thanks to the work of seismologists, we now have more reasonable explanations of the cause of earthquakes. Broadly speaking, earthquakes are caused when deeply buried rocks suddenly break and then snap back into their original position. As noted earlier, the more deeply buried crustal rocks are continually subjected to great stress and strain. As time passes and the pressures build up, the rocks are gradually bent. But despite their ability to bend, if the pressure continues rocks eventually reach their breaking point. It is then that the rocks fracture and suddenly snap back into their original unstrained position. This type of rock fracture is called a *fault*, and the snapping back is known as *elastic rebound*.

Faulting displaces the rocks on either side of the fault. It is this sudden movement of the broken rock that generates the wavelike movements that cause earthquakes. Seismograph records indicate that the tremors associated with the Anchorage quake lasted only about four minutes. Yet it may have taken the subsurface rocks thousands — even millions — of years to finally reach their breaking point.

Seismologists tell us that more than one million earthquakes jar this

48

The arrow points to one side of the fault along which these rocks have been fractured and displaced. (H. E. Malde, U.S. Geological Survey)

planet almost every year. Luckily, only about seven hundred of these yearly jolts are strong enough to do much damage or cause any loss of life. And only a few of these seven hundred will be severe enough to be considered a major quake. Why do not more of these earthquakes cause serious damage? One reason is that most earthquakes occur beneath the sea. Others take place in remote unpopulated areas where they result in little if any loss of life.

Volcanoes and earthquakes have a number of things in common. In fact, many volcanic eruptions are preceded by earthquake activity that warns of the forthcoming eruption. By the same token, faults that cause

49

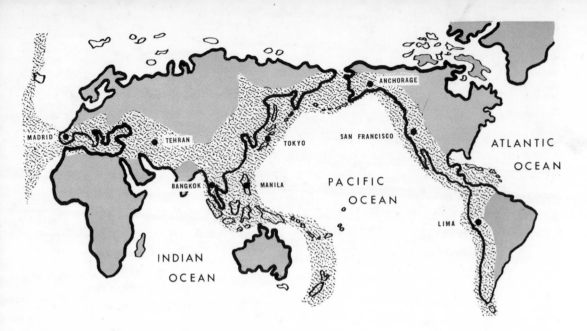

About 80 percent of all of the world's earthquakes occur in the areas shown above.

earthquakes may also create zones of weakness through which magma can reach the surface, thus producing a volcano. It is not surprising, then, to learn that the world's earthquake belts very closely parallel the areas of most recent volcanic activity.

About 80 percent of the world's earthquakes originate in the circum-Pacific belt. This zone of youthful mountain ranges and chains of volcanic mountains borders the Pacific Ocean and coincides with the Pacific Ring of Fire. Alaska is located within this belt, as are such quake-ridden areas as Japan, Mexico, and South and Central America. The Mediterranean and trans-Asiatic belt accounts for roughly 15 percent of the world's earthquakes. This seismic zone reaches from the Atlantic Ocean through the Mediterranean region to southern Asia. Earthquakes scattered throughout the rest of the world account for the remaining 5 percent.

50

The geologic process responsible for earthquakes and crustal deformation is called *diastrophism*. Like volcanism, diastrophism is a great rearranger of the crustal rocks. Diastrophic movements can also greatly affect the landscape. Consider the mighty Sierra Nevada in California. This great mountain range consists of a single block of rock that tore loose in the earth's crust. This massive crustal fragment is more than 400 miles long and ranges from 40 to 80 miles in width. It's highest point — Mount Whitney — lies 14,495 feet above sea level. One can only imagine what a "superquake" this great fault movement must have caused!

10.   The Crust Bends and Buckles

"Solid as a rock." "Unyielding as stone." "Hard as granite." We often use these sayings to emphasize the hardness and stability of objects. And rightly so, for rocks and stone have long been considered the perfect symbols for all that is stable and solid. But are they? Placed in the proper environment, any rock can be squeezed until it will bend and fold like soft plastic.

Proof of such folding can be seen in mountain chains such as the Alps, Rocky Mountains, and Appalachians. In many of these mountains the rocks have been folded into a series of roller-coaster-like highs and lows. These folds, which originated deep within the crust, are now found many thousands of feet above sea level.

Geologists have long puzzled over the origin of folded mountain ranges. They have tried to find out what powerful internal forces could

These beds of sedimentary rocks were originally flat-lying. Crustal disturbances have crumpled them into a series of folds. (U.S. National Park Service)

cause the crust to bend and buckle. Equally baffling is how rock — which is normally hard and brittle — can be forced to bend, and even flow. They have also wondered why most of the folded mountain ranges contain such great thicknesses of sedimentary rock. These rocks, many of which contain fossils of sea animals, may lie as much as three miles above sea level.

Where were such thick layers of ocean sediment deposited and how were these prehistoric sea floors raised into the clouds? Though there are still many missing pieces in the puzzle of folded mountains, some questions have been answered. Geologists tell us that the seashells and ocean sediments were originally laid down on the bottoms of long, sea-filled troughs called *geosynclines*. Over vast periods of time, the streams which emptied into the geosynclines deposited their sediments on the floor of these troughs. Countless plants and animals lived in the now-vanished prehistoric seas. When they died their remains fell to the ocean bottom and were added to the other sediments.

As time passed, the layers of sediments continued to pile up, exerting strong pressure on the strata beneath them. The great weight of the overlying material gradually pressed the deeply buried rock particles tightly together.

Although this pressure was applied very slowly it was continuous. As pressure built up, the rock particles eventually reached a state where they could actually be squeezed along in a sort of "plastic flow." Rocks in this condition might then bend without breaking. As sediments were added to the trough the lower rocks would bend or sag downward. This action deepened the geosyncline and made room for the deposition of still more sediments. The rock record tells us that as much as 35 miles of sediment piled up on the bottom of some of these ancient seaways.

How do we know the seas were there if they vanished millions of years ago? Again, the earth scientist has turned detective. He has gone

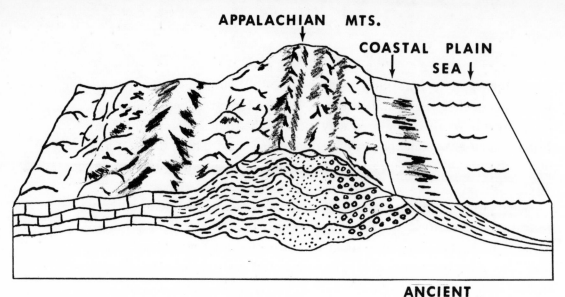

APPALACHIAN MTS.

COASTAL PLAIN

SEA

ANCIENT LANDMASS

MISSISSIPPI VALLEY AREA

APPALACHIAN GEOSYNCLINE

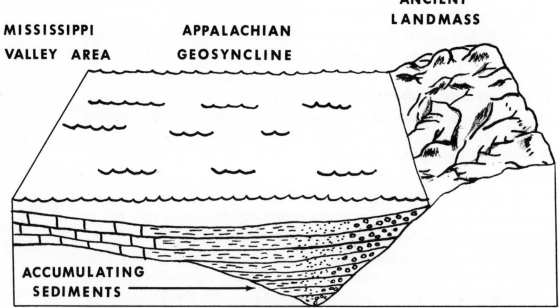

ACCUMULATING SEDIMENTS ⟶

The lower diagram shows the Appalachian Mountain region as it might have appeared when sediments were accumulating in the Appalachian Geosyncline hundreds of millions of years ago.

The upper diagram shows the same region after the sediments had been converted to rocks and uplift occurred.

to the field in search of clues that will reveal something about the geologic history of the area under investigation and he has found evidence of a number of sea-filled troughs recorded in the rocky crust. One of these, the Appalachian Geosyncline, was located about where the Appalachian Mountains lie today. Most of the rocks exposed in this great mountain range were formed from sediments deposited there several hundred million years ago. Later in geologic time, an even larger seaway was located along the site of the Rocky Mountains. Known as the Rocky Mountain Geosyncline, this body of water extended from the Arctic Ocean to the Gulf of Mexico. The sediments laid down in this sinking trough were later changed into the rocks that now make up part of the Rocky Mountains.

This explains, then, how the remains of fossil seashells became locked up in the rocks. It also sheds some light on how and why rock materials may bend or flow. But how were these deformed, fossil-bearing rocks lifted thousands of feet above sea level? This question has baffled geologists for many years — and it still does. Fortunately, geological field studies of folded mountains have told us something about how uplift might have occurred.

To see how these mountains might have formed let us turn back the clock some 80 million years. This was the Age of Dinosaurs, a time when strange reptiles ruled land, sea, and air. Near the end of the reign of the reptiles the bottom of the Rocky Mountain Geosyncline gradually started to rise. As the uplift continued, the water that filled the trough slowly drained back into the deeper ocean basins. Eventually, the sea was no longer there — the ocean bottom was high above sea level.

The forces that raised the old geosyncline also caused folds to develop in the rocks. Finally, the entire Rocky Mountain region was elevated and the tops of the great rock folds became mountains. We can see some of these folds today where they have been exposed by the forces of erosion. But when you see such folds — even in the highest mountains — remember that the rocks were folded deep within the crust.

The uplift and crumpling of the rocks did not occur at once. Instead, this deformation was the result of forces that had been at work within the crust for many millions of years. Yet, in terms of geologic time, these events did take place relatively quickly.

Igneous rocks occur with many of the sedimentary rocks in the Rocky Mountains. These were formed from magma injected into the surrounding rocks during the time of folding and uplift. Faults are also present. Some of the rocks could not endure the pressures of folding and they broke under the strain. As a result, many faults and igneous rock bodies are commonly seen in folded mountain ranges.

However, the most mystifying question is: What mighty force is capable of causing such severe crustal deformation and uplift? To answer this question geologists have developed a number of hypotheses or theories. Although based on much study and research, these are somewhat like scientific guesses. In other words, they are unproved statements made for the purpose of experimentation or testing. Scientists hope, of course, that the experiments and tests will eventually prove that the theory or hypothesis is correct.

The *Contraction Theory* was one of the earliest attempts to explain how folded mountains were formed. This theory assumes that the earth is cooling from an original molten condition. With the cooling of earth's interior, our planet started to shrink, or contract. The contraction gave rise to pressures that crumpled and wrinkled the crust, thus producing folding. We might compare this process to the wrinkling of an apple's skin as its inside shrinks during drying. In this sense, the wrinkles in earth's rocky rind are somewhat similar to the wrinkles in the skin of a dried apple.

As so often happens in science, another explanation proposes just the opposite. The *Expansion Theory* assumes that our earth was originally about one-half of its present diameter of 7,900 miles. In the beginning, this "miniplanet" was uniformly covered by the granitic, continental crust. But with the passage of time the earth began to expand, breaking the crust into blocks that formed the continents. As the earth continued to swell, the crustal blocks were pushed farther apart, and the space between them became the ocean basins.

How could a swelling earth produce folded mountain ranges? When the solid crustal blocks were pushed over the more plastic underlying mantle, a great deal of friction was produced. This dragging effect could cause the rocks to become wrinkled into a series of folds. You can get an idea of how this works if you stand on a rug on a polished floor and slowly slide both feet together. The weight of your body, the movement of your feet, and the friction produced as the rug is dragged over the floor all work together to cause the wrinkles in the rug. This is somewhat similar to what might have taken place between the crust and mantle.

The *Convection Theory* suggests that earth's interior heat produces *convection currents*. Such currents may produce pressure which could cause the rocks to expand and push upward. If you have ever watched a pot of boiling oatmeal, you know that the oatmeal is constantly moving. When heat is applied to the cooler part of the oatmeal, that part sinks because it is more dense than the heated portion. This pushes up the heated part which then loses its heat upon reaching the surface. As the oatmeal that has been pushed up cools, it becomes more dense and gradually settles to the bottom of the pot again. This continual exchange of heat sets up circulating currents of convection cells as long as heat is applied.

The Convection Theory is based on the presence of giant convection cells in the earth's mantle. It is believed that pressure exerted by these convection currents could have caused parts of the crust to move away from each other. This movement would produce drag that might crumple and fold certain parts of the crust.

The heat required to generate convection currents is apparently supplied by the steady decay of radioactive elements. What causes the material in the mantle to expand and flow upward? The rocks of the earth's crust are poor conductors of heat. Consequently, they serve as an insulating blanket that holds in the radioactive heat. This permits a steady buildup of heat which triggers the expansion of the mantle material.

A more recent theory supposes the expansion of the earth as the crust splits along fracture lines. A combination of the expansion and convection theories, this proposal assumes that molten rock material would well up through breaks in the crust. Pressure exerted by the outward flow of melted rock would push the blocks farther apart. As the blocks moved, their leading edges would meet resistance which would crumple the rocks into a series of great folds. This theory assumes that the continents have not always been where they are today. Instead they have "floated" like great crustal barges to their present positions.

Other and more complex theories have also been put forth. But despite all that has been learned, the riddle of folded mountains is still very much of a mystery.

11. Earth's Changing Crust

Earthquakes may rupture the crust and volcanic activity can change and rearrange the materials of which it is made. But as dramatic as these geologic processes are, they are not nearly as important as the slower and more subtle changes that occur on the crustal surface.

The geologic agents that have shaped the face of the crust are so common that they are generally taken for granted. Streams carve their channels and as they do they remove crustal material which will later be deposited as sediment. Mighty rivers of ice plow down mountainsides gouging and scratching the rocks over which they pass. When the glaciers melt, the rocks that they carried will be left behind as evidence of their presence.

In some parts of the world, ocean waves and currents are eating away the shore and redepositing some of this eroded material to form sandbars and beaches elsewhere. Even the mighty mountains do not escape the relentless attack of geologic agents. The most lofty peaks are being steadily lowered as rain and frost wear away the rocks of which they are

Windblown sand has undercut this pedestal of granite in the Chilean desert. (K. Segerstrom, U.S. Geological Survey)

Some earth movements — like the disastrous Madison River Canyon landslide in Montana — move with incredible speed and do great damage. (U.S. Geological Survey)

made. These products of rock weathering may eventually wind up in the valleys below. There they will undergo still more weathering and may eventually be reduced to rich, fertile soil.

The wind is also at work on the surface. Windblown particles of sand act as a natural sandblasting machine to wear away the crustal surface. When the wind dies down, these tiny rock fragments may pile up to form great sand dunes in the desert or along some beach.

In some regions, the effects of gravity may cause large segments of

60

Slumping of the mantlerock and mud slides move relatively slowly, but have still caused much damage in parts of California. (Dr. Gordon B. Oakeshott, California Division of Mines and Geology)

the crust to move downslope. Some of these movements — for example, landslides and rockfalls — occur very suddenly. Others, such as the disastrous mud slides that have plagued parts of California, move much more slowly. The water trapped within the crust can also be an effective geologic agent. Groundwater may pass through soluble rocks and gradually dissolve them. This may produce great underground chambers, or caverns, within the crust. The caves may later be decorated with stalactites and stalagmites formed when the underground water deposits its load of dissolved minerals.

Changes such as those described above are caused by *gradation* of the crust. By means of these gradational processes, the earth materials on

61

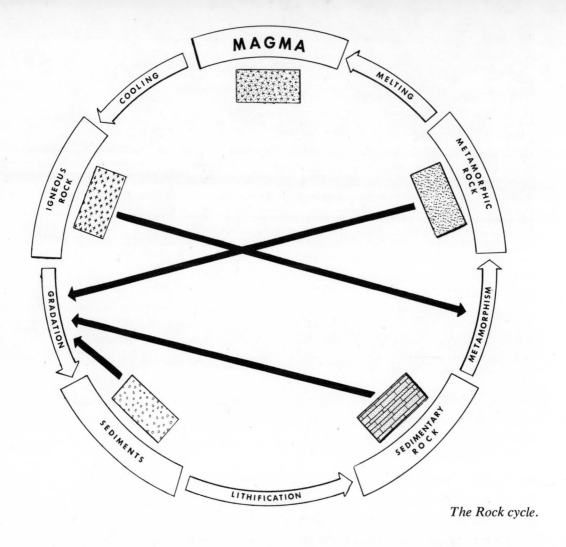

The Rock cycle.

the surface of the crust are constantly being changed and moved about. Gradation, with help from volcanism and diastrophism, is the tool that nature has used to shape the face of the earth and carve the landscapes that we see today.

Gradation, volcanism, and diastrophism not only shape the crust, they are also involved in making the rocks of which it is composed. The rela-

62

tion between the rocks and geologic processes can be seen in the *rock cycle*. This never-ending cycle makes it possible to trace the different paths that rock materials follow and the geologic processes that affect them along the way.

To see how the rock cycle works, let us start with a mass of molten rock and let this magma cool and harden to form an igneous rock. When exposed on the surface, this rock will be eroded and weathered to produce sediments that may be moved and deposited by wind, water, or ice. As time passes, these sediments may be buried under other rock materials and become compacted and cemented together to form sedimentary rock.

Now assume that this deeply buried sedimentary rock, and the remaining igneous rock from which it was derived, are placed under very great heat and pressure. The igneous and sedimentary rocks cannot withstand such severe physical conditions and may become changed into metamorphic rocks. Should the metamorphic rocks continue to be heated and squeezed, they may eventually melt to form magma. When the magma solidifies to form igneous rock, the rock cycle has been completed.

The complete, uninterrupted cycle can be seen on the outer cycle in the diagram on page 62. However, there may be many shortcuts along the cycle and it may never be completed. Some of the shortcuts that bypass major parts of the cycle are indicated by arrows on the diagram.

The rock cycle, like volcanism, diastrophism, and gradation, reminds us that Earth is a most dynamic planet. It has changed much during the 4½ billion years that it has existed and it continues to change even now.

12. A Crustal Jigsaw Puzzle

Many geologists consider the continental crust to be somewhat like the pieces of a gigantic jigsaw puzzle. You can very easily test this jigsaw theory for yourself. First, you will need a map of the world. If you cannot find one in a newspaper or magazine you can trace a map in an atlas. You may want to use the one that appears on page 66 of this chapter. (Be careful not to damage the book.) In addition to the map, you will need some scissors and a roll of transparent tape.

Look at your map and locate the continents, taking care to notice their relation to each other. Do you see a coastline on one continent that might fit the coast of a continent on the other side of the sea? Next, number each of the continental landmasses.

Examine the coastline of each continent very carefully. Does your map indicate the position of the continental shelves (shown as dashed lines on the map on page 66)? If not, outline these submarine platforms on your map, using the map in the book as a guide. This is important, for you will recall that the continental shelves are really the drowned margins of the landmasses. Taken together they make up 11 percent of the continents. When you have numbered the continents and added the continental shelves, carefully cut out the separate landmasses along the dashed lines which represent the margins of the shelves. You now have the pieces for the crustal jigsaw puzzle.

Now see if you can fit the pieces together. Start at an obvious place, like putting the point of South America against the great indentation in the west coast of Africa. You will soon learn that the parts of the continental jigsaw puzzle do not match perfectly. Instead, it is usually necessary to turn and slide them to obtain the most logical fit. More important, you will find that the margins of the submerged continental shelves often

64

When examined carefully, some parts of the continents appear to fit together like a jigsaw puzzle. (Photo by Jennie A. Matthews)

fit more closely than the coastlines proper. This is not surprising, for the shelf areas actually represent the true continental margin of the land-masses.

If you cannot fit all of the pieces into a single supercontinent, try fitting the continents of the Southern Hemisphere together to make one single landmass. If this seems to work do the same for the continents of the Northern Hemisphere.

By now you may be wondering about the reason behind this great crustal jigsaw puzzle. Did the continental crust originally consist of only a single supercontinent, or perhaps two smaller ones? If so, how did each

65

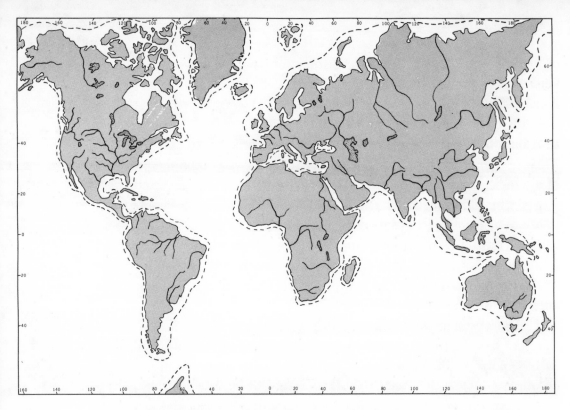

The continental shelves — the submerged margins of the continental landmasses — are shown by means of dashed lines.

of the continents reach their present position? Still more important, what tremendous force could have caused the massive ancestral continent to break up into the smaller chunks of continental crust that we see today? Or is this jigsaw-puzzle fit a mere coincidence that has no scientific meaning whatsoever?

If these questions occur to you, you are not the first to ask them. As early as 1620, Sir Francis Bacon, a famous English philosopher, puzzled over the close resemblance between the Atlantic coastline of South America and that of Africa. Some two hundred years later Antonio Snider also recognized the similarity of certain coastlines on either side

66

of the Atlantic. But he also noticed something else: a close resemblance between certain fossil plants that had been collected in both Europe and America. Antonio Snider may have been the first person to try to work the crustal jigsaw puzzle. In his attempt to determine the geographic distribution of these plant fossils, he tried to fit the continents together. Their approximate match led this early scientist to assume the presence of an ancient master continent.

Then in 1885 an Austrian geologist named Edward Suess noted distinct coastal similarities between the continents of the Southern Hemisphere. He studied the rock formations exposed on these continents and found them to have certain features in common. Consequently, Suess fit the continents together to form one great continent. He named this ancient landmass *Gondwanaland*.

The observations of Snider and Suess were quite significant, for they were not based entirely on the matching of the continents. Instead, they used geologic evidence — fossils and rocks — to support the matching coastlines.

However, it was Alfred Wegener, a German meteorologist, who first attempted to explain the problem of drifting continents. He did this in 1915 when he proposed the Theory of Continental Drift. Wegener suggested the presence of a single massive ancestral continent. He called this monstrous landmass *Pangea* — a word which means "All-Earth."

Wegener assumed that Pangea broke apart about 200 million years ago and the pieces slowly began drifting toward their present positions. To support his idea, Wegener did not rely on the jigsaw puzzle approach alone. He also used rocks and fossils to reconstruct conditions on earth during certain intervals of geologic time. The results of these studies led Wegener to believe that during certain chapters of earth history, the climate, rocks, and life-forms had been quite similar in different parts of the world. These similarities led Wegener to propose that the continents had been closer in the geologic past.

67

In working the crustal jigsaw puzzle you may have found it easier to form two supercontinents rather than one. Recent scientific evidence appears to support the idea of two ancestral continents. The largest, Gondwanaland, is comprised of the Southern Hemisphere landmasses of South America, Africa, India, Australia, Malagasy, Antarctica, and certain submerged fragments of the continental crust. *Laurasia,* located in the Northern Hemisphere, was made up of what is now Eurasia and North America.

Wegener's radical new idea stirred up much controversy among earth scientists of the day. They considered continental drift to be a most outlandish theory, for Wegener could not explain what caused Pangea to break up in the first place. Nor could he explain how his "rafts" of continental crust might have "floated" to their present locations. Thus, as time passed and objections mounted, most scientists either abandoned or scoffed at the idea of drifting continents.

Although ignored or ridiculed for years, continental drift now has the support of most earth scientists. Recent scientific investigations indicate that the continents not only drifted in the past — they are still drifting today. Evidence to support this belief comes from many sources. Geologic studies of the great mountain ranges on the western borders of North and South America provide one clue. They might have been formed when the rocks crumpled and piled up as the leading edge of the continents drifted westward. It has also been learned that certain rocks on opposite sides of the Atlantic Ocean have a similar origin and are rather close in age. These rocks have also been exposed to the same geologic processes — for example, glaciation — during the same portions of geologic time.

The study of fossils has added yet another vital link in the chain of evidence to support continental drift. The presence of certain plants that occur in glacier-deposited sediments that cover parts of Africa, South

America, Australia, New Zealand, and India is especially significant. Most plant scientists agree that these ancient plants could not have spread across the broad stretches of the open sea. Instead, these now widely separated areas must have once been joined together.

The fossil remains of a meat-eating reptile found on opposite sides of the Atlantic in rocks of the same age is another clue. This freshwater creature could not possibly have migrated across the open sea. More recently, the jawbone of an ancient freshwater amphibian and the bones of a reptile were discovered in Antarctica. Similar remains had been found earlier in Australia and South Africa. Here again is strong evidence suggesting that the continents of the Southern Hemisphere may have originally been united.

Despite the mounting evidence in support of drifting continents, much remains to be learned. The force responsible for causing the ancestral supercontinents to break up is still not known and there are other mysteries as well. But evidence continues to pile up, and hopefully we may have a final solution to this fascinating puzzle in the relatively near future.

13. Is the Oceanic Crust Spreading?

Strange as it may seem, strong evidence to support drifting of the landmasses has been found on the floors of the oceans. This evidence has come to light as oceanographers have studied the sea bottom in order to learn more about the nature of the oceanic crust and its relation to the continental crust.

69

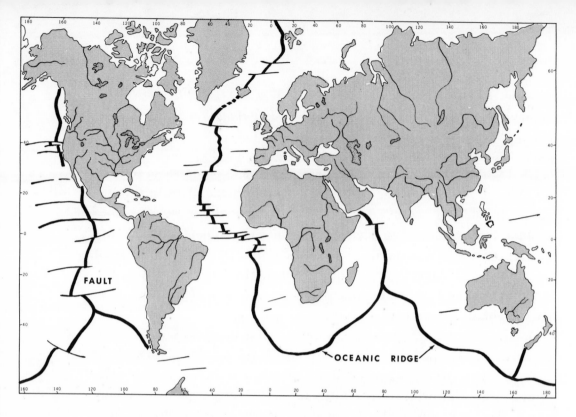

Map showing oceanic ridges and the offsetting system of faults.

Marine geologists have been particularly interested in the oceanic ridges which form a rugged undersea mountain system some 40,000 miles long. One of these, the Mid-Atlantic Ridge, is of special interest to supporters of the continental drift theory.

On the map shown above, you will see that this great submarine mountain chain runs the length of the Atlantic Ocean. Notice, too, how its path roughly follows the outline of the continental margins. There is evidence to suggest that the formation of this extensive undersea range marks the point of separation of the continents on either side of it.

Scientists have learned much about the crust and continental drift by

70

studying the rocks on and adjacent to the Mid-Atlantic Ridge. Their findings indicate that the oceanic crust is not stationary but is slowly expanding. This spreading appears to be associated with volcanic activity that originates from the deep rift, or trench, along the spine of this mid-ocean ridge.

Molten rock has apparently boiled out of the crust along the rift at many times in the geologic past. Each time this has happened, the solid rock on either side of this great crack has been pushed progressively farther away. And, because the basaltic oceanic crust is rather thin and brittle, any movement along the ridge causes it to break rather easily. Thus, the Mid-Atlantic Ridge is intersected by many faults. Most of these fractures have a lateral or sideways movement.

Although the ocean floor appears to be tearing apart along the rift, this ever-widening crack does not extend downward indefinitely. Instead, it seems to be continually filling up with lava rising from within the crust. The most recent lava deposits are found in and along each side of the rift, and the lava flows become increasingly older on either side of it. This banded pattern of progressively older lava beds suggests that the sea floor is continually spreading and being rebuilt by molten rock pouring out of the central rift.

The concept of sea-floor spreading is of great interest to supporters of the continental drift theory. They have suggested that as the oceanic crust of the sea floor has spread, it has carried along with it the lighter rocks of the continental crust. If this is true, a spreading ocean bottom might have served as a great "conveyor belt" that has gradually carried the continents to their present positions.

How is it known that the rocks of the ocean floor become older as we move away from the Mid-Atlantic Ridge? This has been determined by studying the effects of the earth's magnetic field on the rocks. When sediments containing particles of magnetic iron minerals were deposited, the

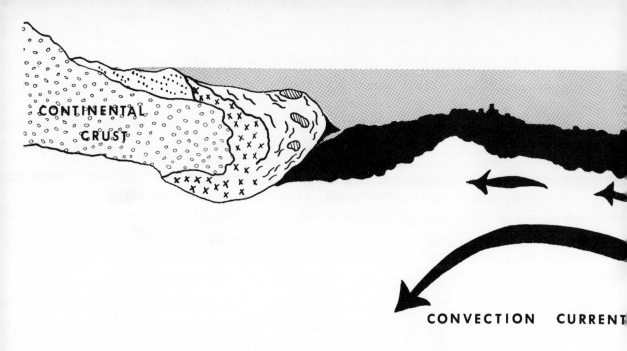

CONTINENTAL CRUST

CONVECTION CURRENT

Cross section showing how the oceanic crust is being torn apart to form the mid-ocean ridges.

fragments of iron lined up with the earth's magnetic field. In other words, each particle of iron becomes a tiny "compass needle" pointing toward earth's magnetic poles. Fortunately, these minerals retain their magnetic properties for hundreds of millions of years. This rock characteristic is known as *paleomagnetism* (literally "ancient-magnetism").

In very young rocks the little compasses are in line with the direction of the magnetic field of the earth at that time. But in older rocks, the magnetic mineral grains may point in quite different directions from

72

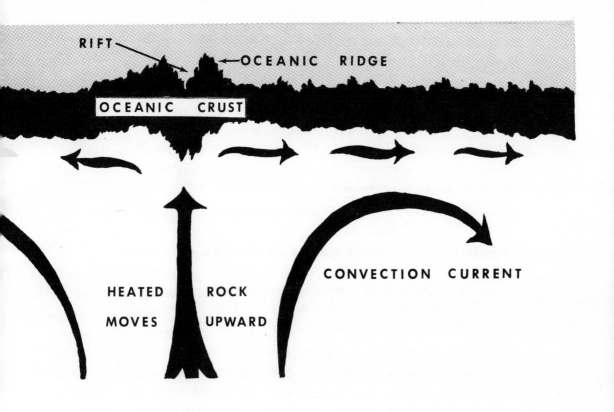

RIFT
OCEANIC RIDGE
OCEANIC CRUST
CONVECTION CURRENT
HEATED ROCK
MOVES UPWARD

our present magnetic poles. This variation suggests that the earth's magnetic field has changed many times in the history of the earth. In fact, in times past the North and South Poles have probably been reversed.

What do wandering poles and paleomagnetism have to do with drifting continents? Geophysicists have studied these phenomena and constructed polar wandering curves. They assume that if the landmasses have always been where they are today, all of the continents would yield the same polar curves. On the other hand, if the continents have shifted, the study of fossil magnetism would produce different polar curves for the different continents. The curves constructed by many earth scientists

73

strongly suggest that the continents have changed positions during the geologic past.

Paleomagnetic studies of the lava flows on either side of the Mid-Atlantic Ridge reveal that one flow will have normal polarity and that the next will show reversed polarity. In other words, one flow was extruded onto the ocean floor when our magnetic pole was much as it is today. But the next band of lava was deposited hundreds of thousands of years later when the poles were reversed. This has resulted in the ocean floor becoming magnetized in a stripelike pattern of alternating bands of normal and reversed polarity. These alternating bands may have recorded the growth of the ocean floor in much the same way that the rings in a tree trunk will reveal the growth of the tree. Some scientists view these paleomagnetic records as the "footprints" of the continents as they moved to their present position.

Curiously enough, additional evidence of continental drift and its relation to sea-floor spreading may soon come from outer space. Among the pieces of scientific equipment left on the moon by Apollo 11 astronauts was a special reflecting prism. Astronomers have directed laser beams toward this prism and determined the precise distance from the earth to the moon. Measurements by the laser-prism method will be checked regularly to see if the distance between the laser and prism is increasing. If it does increase this will be further proof that the continents are drifting now and probably have been for a very long time.

14. The Crust Reveals Earth's History

Most of us are fascinated by objects that are very old. Old coins, stamps, furniture, and even antique automobiles are treasured by many collectors. And, generally speaking, the older such objects are the more interesting they become. The same thing can be said of the earth. This remarkable planet is extremely ancient and its early history is especially interesting.

How do we know the age of the earth and how can we trace its history? Earth scientists have developed techniques of studying the various rock layers of the earth's crust. They are able to determine much of what has happened to the earth and its inhabitants during the past 4½ billion years.

Almost every rock in the crust adds a bit of information to the fascinating story. For example, each rock yields some evidence of the conditions under which it was originally formed. Rocks may also contain some indication of the changes that they have undergone since they were formed. The more fiery episodes in earth history are revealed by the igneous rocks. They indicate times of great volcanic explosions, of showers of glowing cinders and ash, and red-hot lava flows. The metamorphic rocks also provide evidence of some of earth's more violent events. These changed rocks are clues to times of great crustal unrest. They tell of a buckling crust, of minerals that were squeezed and melted, and of new rocks formed from old.

However, most of what we know about the history of the earth and its various life-forms has been gleaned from the study of sedimentary rocks. Many of these rocks are veritable treasure troves of earth history. They have recorded the former presence of mighty glaciers, restless winds, and shifting sands, and of eroded lands and flooding rivers. They

The rock layers exposed in Petrified Forest National Park are like rocky pages of earth history to the geologist. (Fred Harvey Company)

may also contain clues that help to reconstruct the distribution of the lands and seas of prehistoric time. Some sedimentary rocks contain fossils. These remains or traces of prehistoric life are particularly valuable, for they reveal the development of life on earth.

The earliest chapters in earth history are found in the oldest known rocks of the earth's crust. Formed at the very dawn of geologic time,

76

these ancient rocks are usually buried deep within the crust. And —
unless they have been exposed by uplift and erosion — they are usually
covered by younger rocks which have been deposited on top of them.
These younger strata have documented the more recent events in earth
history.

Unfortunately, this earthen history book is not always easy to read.
Crustal deformation has caused certain of the rocky "pages" to become
shuffled out of order or to be lost forever. Worse yet, many rock layers
have been charred and crumpled by violent episodes of volcanic activity
and mountain building. Still others are so deeply buried in the crust that
they are not available for study.

Despite these missing pieces in the puzzle of earth history, much has
been learned about this ancient planet. Like a detective gathering clues
to solve a crime, the historical geologist searches the rocks for evidence
of past events on earth. Luckily, the work of the geological sleuth has
not been in vain, for he has used bits of evidence from many sources to
unravel the ancient mysteries of geologic time.

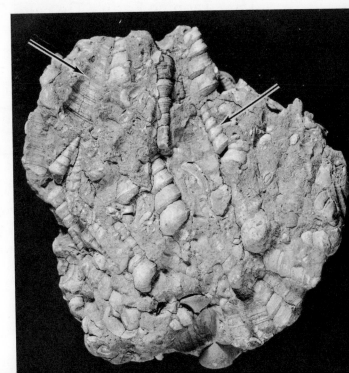

Fossils — the traces or remains of organisms that lived during prehistoric time — have revealed much about the history of life on earth. The arrow at upper left points to the imprint of a snail shell while the arrow on the right indicates the hardened mud filling of a similar shell. (Smithsonian Institution)

Much of what we know about the geologic past has come from the study of fossils. These preserved bones, teeth, shells, and tracks have been especially valuable in understanding life-forms of the geologic past. *Paleontologists* (experts on fossils) have found that each fossilized plant or animal provides some clue as to where it lived, how it lived, and when it lived. Fossils also provide valuable clues that permit us to reconstruct the geography and climate of prehistoric time.

Investigations of the oldest known fossils suggest that life began on this planet some 3 billion years ago. Paleontologists have traced the parade of life from these first simple plantlike organisms through the Age of Dinosaurs all the way to ancestral man.

15. Tapping Earth's Treasure Chest

High on a windswept mesa in Utah, a prospector grins as he listens to the welcome ticking of his Geiger counter.

A diver off the coast of California hovers above the ocean floor carefully collecting samples of rounded masses of manganese.

On Alaska's frozen north slope, a drilling rig bores relentlessly through the earth's crust. Nearby, a geologist hunches over his microscope carefully examining rock samples from the borehole. He is looking for geological evidence that may lead to another great oil discovery.

Uranium prospectors use Geiger counters and similar instruments to detect radioactive minerals in the rocks. (Union Carbide Corporation)

Left: The marine geologist studies the ocean floor in hope of undercovering valuable undersea deposits of minerals. (Naval Undersea Research and Development Center)
Right: This drilling rig is in Cook Inlet, Alaska. This is but one of the many places that geologists search for oil. (Standard Oil Company of California)

Despite their different activities and their wide geographic distribution, the aforementioned men all have one thing in common: They are geologists looking for clues to the great mineral wealth locked up in the crustal rocks.

Man has always relied on the earth's crust. During prehistoric time ancient man used stones to fashion crude weapons and tools. He also found shelter in caves dissolved from the rocks. Today's modern industry and technology are even more dependent on the crust. Natural resources extracted from rocks and minerals are the raw materials that are basic to our economy. And coal and petroleum — products of the remains of prehistoric plants and animals — furnish the necessary energy to convert these raw materials to useful products.

We utilize geologic information and geologic products in many ways. For this reason the geologist plays an increasingly important role in our lives. It is his duty to provide modern civilization with the mineral fuels, ores, and other economic minerals so vital to industrial growth.

What are the earth materials that have had and continue to have such a pronounced effect on our lives? How were they formed and how are they found and extracted from the crust? These are but a few of the questions that the geologist must answer as he looks for the natural mineral riches in the crustal rocks.

Fossil fuels — coal, oil, and natural gas — are among the most sought-after treasures in the crust. These valuable energy-producers are called fossil fuels because they are formed from the remains of ancient plants and animals.

Coal, a fossil fuel of plant origin, is common in certain sedimentary rocks. Although coal is composed mostly of carbon, hydrogen, oxygen, and nitrogen, sulfur and other elements may be present as impurities. Coal is formed by *carbonization*, a process whereby decaying plant material loses water and gases with a resulting concentration of carbon.

80

Within recent years, petroleum has replaced coal as a source of heat and energy in many areas. Even so, coal is quite vital to certain industries and is still the most important solid fuel in the world.

Although man has used oil since the earliest times, the first commercial oil well was not drilled until 1859. But since the day that Colonel Edwin L. Drake drilled that historic well in Titusville, Pennsylvania, this liquid "black gold" has become one of the world's most important resources. Most geologists believe that petroleum has formed from the remains of microscopic plants and animals that lived in prehistoric seas. These organic remains were buried in the mud and sand on the floor of these ancient oceans and gradually decomposed. Scientists do not yet understand exactly how these plant and animal products were transformed into petroleum. It seems, however, that the process requires great lengths of time, accompanied by increases of temperature and the compression of sediments.

Petroleum is seldom found in the rocks in which it formed. Instead, the oil and gas migrate out of the source rock into more porous rocks that act as natural underground pipelines to move them through the crust. This movement normally continues until the petroleum reaches an area where the structure of the rocks is such that the oil can no longer flow. The oil is then trapped where it may eventually form large pools.

The petroleum geologist is continually searching for oil traps in the earth's crust. He knows that these are most likely to occur in areas where the rocks have been deformed. Thus, those areas where the subsurface rocks have been folded and faulted might be favorable for the accumulation of commercial quantities of oil and gas.

As noted earlier, the rocks of the earth's crust also contain many different kinds of metallic minerals. Because of their great usefulness to man, metals are among the most valuable of our mineral resources. Known also as ore minerals, the metallic minerals include such widely

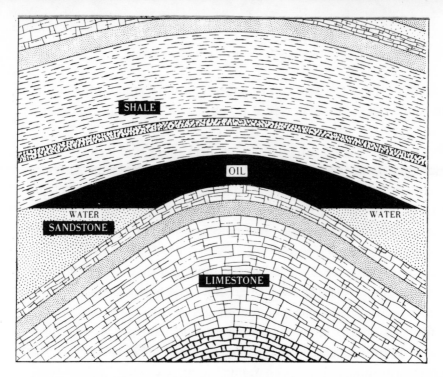

Petroleum may be trapped in the subsurface where certain crustal rocks have been deformed. (Arabian American Oil Company)

used substances as iron, copper, aluminum, tin, zinc, lead, silver, and gold. Also important are the *radioactive minerals* such as uraninite (or pitchblende) and carnotite, for they are sources of nuclear energy.

Ore minerals may be found in either igneous, sedimentary, or metamorphic rocks. Many minerals occur in *veins*. Those veins associated with igneous rocks may develop when mineral-bearing magma is injected into the surrounding rocks. Others are formed when circulating groundwater picks up metallic compounds and deposits them in crevices in the rock.

Some mineral deposits are the results of weathering of the crust. For example, bauxite (a valuable ore of aluminum) may be formed as the result of the chemical weathering of clays and granites of high aluminum content. Some of the larger deposits of iron ore are thought to have been formed in much the same way.

In some places ore minerals are found in *placer deposits*. These are mineral-bearing accumulations of sand and gravel in the beds of streams that eroded the rocks in which the minerals were originally formed. The famous California gold discovery of 1849 was made by a miner panning a placer deposit in the bed of the Sacramento River.

Although not as glamorous as petroleum or ore minerals, the non-metallic minerals are very important. These rocks and minerals are widely used in industry, and for this reason they are also called the industrial minerals. Some of these minerals such as halite are vital to our health and well-being, and most of them play an important role in our everyday lives. Sulfur is used to make paper, sulfuric acid, and gunpowder, and gypsum is an important constituent in plaster of paris, plaster, paint, and wallboard. Also important are such widely used industrial rocks and minerals as asbestos, sand, clay, limestone, and mineral fertilizers.

The earth's crust, then, is not only interesting, it is the foundation of our very existence. Without its soils and mineral resources we could not live as we do today. This is why the geologist, working in many parts of the world, continually probes the earth's crust. He knows that the crustal rocks contain the keys to a better understanding of the earth and its development through time. But more important, he realizes that he must continue to find new ways to discover and conserve our mineral resources if we are to survive on the fragile planet that is our home.

Glossary

ATMOSPHERE — The air surrounding the earth.

BEDROCK — The unweathered solid rock of the earth's crust.

BIOSPHERE — The living realm of the earth's plants and animals.

CARBONIZATION — The process whereby organic remains are reduced to carbon or coal.

CONTINENTAL DRIFT THEORY — A theory that all continents were at one time a single landmass that broke up and drifted apart.

CONTINENTAL SHELF — Relatively shallow ocean floor bordering a continental landmass.

CONVECTION — A mechanism by which material circulates because its density differs from that of surrounding material. The density differences are commonly the result of heating.

CONVECTION CURRENT — A closed circulation of material sometimes developed during convection.

CORE — The dense innermost zone of the earth that is surrounded by the mantle.

CRUST — The outer shell of the solid earth surrounding the mantle.

DIASTROPHISM — Movements within the rocky crust of the earth.

EARTHQUAKE — The shaking of the ground as a result of movements within the earth, most commonly associated with movement along faults.

EPICENTER — Point of the earth's surface directly above the focus or hypocenter of an earthquake.

FAULT — A fracture in a rock surface, along which there is displacement of the broken surfaces.

FOCUS — *See* Hypocenter.

84

FOSSIL — The remains or traces of prehistoric organisms buried by natural causes and preserved in the earth's crust.

FOSSIL FUELS — Organic remains used to produce heat or energy by combustion; examples are coal, oil, and natural gas.

GEOLOGIST — A scientist who studies the earth.

GEOLOGY — Study of the earth.

GEOSYNCLINE — A great downward flexure of the earth's crust, usually tens of miles wide and hundreds of miles long.

GLACIER — A slowly moving mass of recrystallized ice flowing forward as a result of gravitational attraction.

GROUNDWATER — Water which penetrates into spaces within the rocks of the earth's crust.

HYDROSPHERE — All of the water upon the earth's surface or in the open spaces below the surface.

HYPOCENTER — The point of origin in the earth's crust or mantle where an earthquake occurs.

IGNEOUS ROCK — Rocks which have solidified from lava or molten rock called magma.

LITHOSPHERE — The entire solid part of the earth (crust, mantle, core).

MAGMA — Molten rock material in a liquid or pasty state, originating within the earth at a high temperature.

MANTLE — The thick, dense part of the lithosphere beneath the crust and to a depth of about 1,800 miles below the surface.

MANTLEROCK — The layers of loose weathered rock lying over solid bedrock.

METAMORPHIC ROCKS — Rocks that have been changed from their original form by great heat and pressure.

MID-OCEAN RIDGE — Submarine mountain range lying along the centerline of most ocean basins.

85

MINERAL — A naturally occurring inorganic substance possessing definite chemical and physical properties.

MOHO — *See* Mohorovičić discontinuity.

MOHOROVICIC DISCONTINUITY — The zone of contact between the crust and the mantle.

OCEANOGRAPHY — The scientific study of the sea and its characteristics.

PALEOMAGNETISM — The study of the earth's magnetic field as it has existed during geologic time.

PALEONTOLOGIST — A scientist who studies fossils.

PALEONTOLOGY — The science which deals with the study of fossils.

ROCK — Any naturally formed mass of mineral matter that makes up an essential part of the earth's crust.

SEDIMENT — Rock fragments that have been deposited by settling from a transportation agent such as water, ice, or air.

SEDIMENTARY ROCK — Rocks formed by the accumulation of sediments.

SEISMIC WAVES — Earthquake waves.

SEISMOGRAM — The record made by a seismograph.

SEISMOGRAPH — An apparatus to register the shocks of earthquakes.

SEISMOLOGIST — One who studies and interprets the effects of earthquake activity.

SEISMOLOGY — A geophysical science which deals with the study of earthquakes.

SPECIFIC GRAVITY — A number that compares the density of a substance with that of water.

STRATA — Rock layers or beds.

STRATIFICATION — Bedding or layering in sedimentary rock.

TOPOGRAPHY — The physical features or configuration of a land surface.

86

VOLCANO — The vent from which molten rock materials reach the surface, together with the accumulations of volcanic materials deposited around the vent.

VOLCANOLOGIST — One who studies and interprets volcanoes and their related phenomena.

WEATHERING — The natural physical and chemical breakdown of rocks under atmospheric conditions.

Index

Alaska, 46, 50
Alps, 30, 51
Aluminum, 10, 82
Anchorage, Alaska, 1964 earthquakes at, 46-47, 48
Animals, 8-9, 14, 20
  fossils as evidence of supercontinent, 69
Apollo space flights, 28, 35, 74
Appalachian Geosyncline, 55
Appalachians, 51, 55
Asbestos, 83
Ash, volcanic, 42
Atlantic Ocean, 31, 32, 46, 50, 68, 70, 74
Atmosphere, 4, 5, 84
Atmospheric weathering, 5
Atoms, 9

Bacon, Sir Francis, 66
Basalt, 26-27, 71
Bauxite, 82
Bedrock, 14, 16-17, 20, 84
Biosphere, 8-9, 84
Body (seismic) waves, 24
Boulders, 14, 20
Brick, 13

Calcium, 11
Carbon dioxide, 5
Carbonization, 80, 84
Carlsbad Caverns, N.M., 33
Carnotite, 82
Caves, 33, 61
Chemical compounds, 10, 11
Chemical elements. *See* Elements
Chemical weathering, 82
Cinders, volcanic, 42
Clay, 13, 20, 82, 83
Coal, 8, 20, 21, 80-81
Cobalt, in earth's core, 37
Columbia River Plateau, 44
Composition of crust, 22, 26
Compounds, chemical, 10, 11
Compression (seismic) waves, 23-24
Continental crust, 26, 27, 28, 30, 64, 69
  composition of, 26, 27, 28
  thickness of, 27

Continental Drift, Theory of, 67-69, 70, 71, 74, 84
Continental shelves, 30, 64-65, 84
Continents, 28-30, 32, 57, 58, 64-69
  area of, 28, 30
  average height above ocean floor, 28
  average height above sea level, 30
  distribution of, 30
  highest point, 30
  percentage of earth surface covered by, 28-30
Contraction Theory, 56
Convection current, 57-58, 84
Convection Theory, 57-58
Copper, 10, 12, 82
Core, earth's, 37, 40, 84
Crust
  borings into, 23, 34
  composition of, 22, 26
  defined, 84
  origin of term, 22
  seismological study of, 23-26
  structural weaknesses, 44-46, 48, 50
  thickness of, 3, 26, 27
Crustal deformation, 21, 48, 51, 56, 75, 77
Crystals, 11

Diamond, 10
Diastrophism, 51, 62, 63, 84
Dinosaurs, Age of, 55
Drake, Edwin L., 81
Dunes, 60
Dust, volcanic, 42

Earth
  age of, 63, 75
  ancient theories of phenomena of, 33-34, 39, 47-48
  formation theory, former, 22
  history of, studies, 75-78
  history of, theories, 22, 56-58
  interior of, 33-37
  interior pressure, 34
  interior temperature, 34
  magnetic field of, 34, 71-73
  radius of, 3
  space photographs of, 28-30

Earth core, 37, 40
Earth mantle, 35-36, 37, 40, 41
   convection cells in, 57-58
Earthquake belts (zones), 44, 50
Earthquakes, 47-51, 84
   ancient theories of, 33-34, 47-48
   annual number of, 48-49
   cause of, 48
   epicenter of, 25, 84
   focus of, 23, 25
   hypocenter of, 23, 25, 85
   seismic waves of, 23-25
   and volcanoes, 42, 49-50
Elastic rebound, 48
Elemental state minerals, 10
Elements, chemical, 9-10, 11
   most abundant mineral-making, 10-11
   radioactive, 42, 58
Epicenter of earthquake, 25, 84
Erosion, 5-8, 14, 16, 19, 20, 56, 59-61, 63
Everest, Mount, 30, 32
Expansion Theory, 57, 58
Extrusive rocks, 18

Faults, 48, 49, 56, 71, 84
"Floating" of crust, 27, 58, 67-68
Folded mountain ranges, 51-56
Folding, theories on, 56-58
Fossil fuels, 80-81, 85
Fossils, 9, 20, 53, 55, 85
   as evidence for continental drift, 67, 68-69
   oldest known, 78
   study of, 76, 78
Frost, 59
Fuels, 80-81
Fujiyama, Mount, 43

Gas, natural, 80, 81
Gases, 4, 5
Geologists, 3, 9, 14, 17, 80, 81, 85
   historical, 75-78
Geosynclines, 53-55, 85
Glaciers, 20, 59, 68, 85
Glass, 13
Gold, 10, 12, 18, 82, 83
Gondwanaland, 67, 68
Gradation, 61-62, 63
Grand Canyon, Ariz., 32
Granite, 13, 18, 19-20, 28, 82

Granitic rocks of upper crust, 26, 27
Gravel, 14, 20, 83
Groundwater, 5, 61, 82, 85
Gypsum, 12, 20, 83

Halite, 10, 20, 83
Hawaii Volcano Observatory, 44
Hawaiian Islands, 43, 44, 46
Heat, internal earth, 41
   at center, 34
   crustal, causes of, 21, 42, 58
Himalayas, 30
Hydrosphere, 4, 5, 30, 85
Hypocenter of earthquake, 23, 25, 85

Igneous rocks, 17-20, 21, 26, 44, 56, 75
   defined, 17, 85
   ores in, 82
   in rock cycle, 63
   types of, 18
Industrial minerals, 12, 83
Inner core of earth, 37
"Inner space," 33-37
Inorganic substances, 11
Intrusive rocks, 18-20, 21
Iron, 11, 12, 82, 83
   in earth's core, 37

Kilauea Volcano, 44

L (long) waves, 24
Lakes, 5
Landmasses. See Continents
Landscape formation, 59-63
Landslides, 61
Laurasia, 68
Lava, 18, 22, 39, 42, 43-44
   deposits on ocean floor, 71, 74
Lead, 12, 82
Life, 4, 8-9, 14, 75, 76, 78. See also Animals;
      Fossils; Plants
Limestone, 20, 21, 83
Liquids, 4
Lithification, 16
Lithosphere, 4, 9, 22, 85
Lower crust, 26

Magma, 17, 39-42, 50, 56, 82
   defined, 85
   in rock cycle, 63

Magma reservoirs, 40
Magnesium, 11
Mantle, earth's, 35-36, 37, 40, 41
  convection cells in, 56-57
  defined, 85
Mantlerock, 14, 16, 85
Marble, 21
Matter
  states of, 4
  zones (spheres) of, 4
Metallic minerals, 11-12, 18, 81-83
Metals, 9, 12, 81, 82
Metamorphic rocks, 17, 21, 75
  defined, 21, 85
  ores in, 82
  in rock cycle, 63
Meteorites, 5, 34, 37
Meteors, 5
Mid-Atlantic Ridge, 31, 70-71, 74
Mindanao Trench, 32
Mine, deepest, 23
Mineral resources, 3, 9, 14, 20, 21, 80-83
Mineralogy, 11
Minerals, 9-12, 22
  basic groups of, 11-12
  as basic ingredients of rocks, 9, 14
  defined, 11, 86
  in elemental state, 10
  fuels, 80-81
  industrial, 12, 83
  metallic, 11-12, 18, 81-83
  most abundant elements in, 10-11
  in native state, 10
  nonmetallic,. 11-12, 20, 83
  number of, 11
  radioactive, 42, 82
  rock-forming, 12
Moho, 25-26, 35
Mohorovičić, Andrija, 25
Mohorovičić discontinuity, 25-26, 86
Molten rock. See Magma
Mountain building, 21, 42, 77
  by faulting, 51
  by folding, 51-58
Mountain ranges, 30
  continental drift and, 68
  faulted (Sierra Nevada), 51
  folded, 51-56
  thickness of crust under, 27
  underwater, 31, 70

Mountains
  highest, 30
  volcanic, 42-44. See also Volcanoes
Mud, 20
Mud slides, 61

Native state minerals, 10
Natural gas, 80, 81
Nickel, in earth's core, 37
Nitrogen, 5
Nonmetallic minerals, 11-12, 20, 83
Northern Hemisphere
  ancestral supercontinent of, 68
  land vs. water area, 30

Ocean, 9, 57
  area of, 30
  average depth of, 5
  deepest point, 32
  distribution of, 30
  as erosive force, 59
  percentage of earth surface covered by, 5, 30
Ocean floor, 30-32, 69-74
  mapping of, 30-31
  paleomagnetic studies of lava on, 74
  ridges, 31, 32, 70
  sediments, 5, 30, 53, 55, 71-72
  spreading of, 71, 74
  trenches, 31-32, 71
  volcanism, 43, 46, 71
Oceanic crust, 26-27, 28, 30, 69-71
  composition of, 26-27, 31
  spreading of, 71, 74
  thickness of, 27
Oceanography, 86
Oil. See Petroleum
Ore, 12, 80
Ore minerals, 81-83. See also Metallic minerals
Organic (plant and animal) matter, 8-9, 14
Outer core of earth, 37, 40
Oxygen
  abundance in minerals, 10-11
  in air, 5

P (primary) waves, 24, 35, 37, 41
Pacific Ocean, 32, 43, 46-47
  volcanic belts, 44-46, 50
Pacific Ring of Fire, 46, 50
Paleomagnetism, 72, 73, 74, 86

Paleontology, 78, 86
Pangea, 67, 68
Pelée, Mount, 38-39
Petroleum, 8, 21, 80, 81
    first well, 81
Pikes Peak, 19-20
Pitchblende, 82
Placer deposits, 83
Plants, 8-9, 14, 20
    fossils as evidence of supercontinent, 67, 68-69
Plutonic rocks, 18
Poles, magnetic, 72-73
Potassium, 11
Pressure, internal earth, 41, 48
    at center, 34
    crustal, causes of, 21, 42, 56, 57
Puerto Rico Trench, 32
Push (seismic) waves, 23-24

Radioactive elements, 42, 58
Radioactive minerals, 42, 82
Rain, 5, 59
Rainier, Mount, 43
Ridges, ocean, 31, 32, 70
Rivers. *See* Streams
Rock cycle, 63
Rockfalls, 61
Rock-forming environments, 17
Rocks, 3, 4, 9, 12-22
    basaltic, 26-27
    basic classes of, 17
    bed *vs.* mantlerock, 14-16
    defined, 86
    differences in, 14
    as evidence for continental drift, 67, 68
    extrusive, 18
    granitic, 26, 27
    igneous, 17-20, 21, 26, 44, 56, 63, 75, 82
    intrusive, 18-20, 21
    lower crust, 26
    metamorphic, 17, 21, 63, 75, 82
    molten (magma), 17, 39-42, 63, 71, 82
    plutonic, 18
    sedimentary, 17, 20-21, 53-55, 56, 63, 75-76, 80, 82
    stratified, 20
    underground, seismological study of, 23-26
    upper-crust, 26
    volcanic, 18, 26, 39-42

Rock salt, 20. *See also* Halite
Rocky Mountains, 30, 51, 55-56
Rocky Mountain Geosyncline, 55

S (secondary) waves, 24, 35, 37, 41
St. Pierre, Martinique, 37-39
Salt, 10, 12, 21
Sand, 13, 14, 16, 20, 83
Sand dunes, 60
Sandstone, 13, 16, 20
Sea. *See* Ocean
Seashell fossils, in mountain ranges, 53, 55
Sediment, 5, 20, 59, 63
    defined, 86
    ocean floor, 5, 30, 53, 55, 71-72
Sedimentary rocks, 17, 20-21, 53-55, 56
    defined, 20, 86
    mineral resources in, 80, 82
    as record of earth history, 75-76
    in rock cycle, 63
Seismic belts, 44, 50
Seismic waves, 23-25, 35-37, 41, 46-47
Seismograms, 23, 24, 86
Seismographs, 23, 24, 25, 86
Seismology, 23-26, 35, 48, 86
Shale, 20
Shasta, Mount, 43
Shear (seismic) waves, 24
Shells, 20, 53, 55
Sierra Nevada, 51
Silicon
    abundance in minerals, 10-11
    in earth's core, 37
Silt, 20
Silver, 18, 82
Snider, Antonio, 66-67
Snow, 5
Sodium, 11
Sodium chloride, 10
Soil, 3, 4, 9, 14, 20, 60
Solids, 4
Southern Hemisphere
    ancestral supercontinent of, 68-69
    land *vs.* water area, 30
    matching of continental coastlines, 65-67
Spheres of matter, 4
Stalactites, 61
Stalagmites, 61
Strata, 20, 86
Stratification, 20, 86

91

Streams, 5, 20, 59
  placer deposits in, 83
Suess, Edward, 67
Sulfur, 10, 12
Supercontinents, 67-69

Thickness of crust; 26
  average, 3
  continental, 27
  oceanic, 27
Tile, 13
Tin, 12, 82
Topography, 32, 86
Transverse (seismic) waves, 24
Trenches, ocean, 31-32, 71

Underground caverns, 33, 61
Underground water, 5, 61, 82
Upper crust, 26
Uranite, 82

Volcanic belts (zones), 44-46, 50
Volcanic debris, 42

Volcanic rocks, 18, 26, 39-42
Volcanism, 39-46, 51, 62, 63
  ocean floor, 43, 46, 71
Volcanoes, 22, 33, 37-46, 75, 77
  ancient theories of, 33, 39
  causes of, 39-42
  defined, 87
  earthquakes and, 42, 49-50
  location of, 44-46
  number of active, 44
  origin of word, 39
  shapes of, 42-43
Volcanology, 40-46, 87

Water, 4, 5
Water erosion, 5, 19, 20, 59
Weathering, 5, 8, 14, 16, 19, 59-60, 63, 82
  defined, 87
Wegener, Alfred, 67-68
Whitney, Mount, 51
Wind erosion, 19, 20, 60

Zinc, 82